Environmental, Groundwater, and Engineering Geology

Environmental, Groundwater, and Engineering Geology

Contributors

Bizhan Shirdel et al.

AURIS
Reference

www.aurisreference.com

Environmental, Groundwater, and Engineering Geology

Contributors: Bizhan Shirdel et al.

Published by Auris Reference Limited

www.aurisreference.com

United Kingdom

Environmental, Groundwater, and Engineering Geology

ISBN: 978-1-78154-983-4

British Library Cataloguing in Publication Data
A CIP record for this book is available from the British Library

Printed in the United Kingdom

Exclusively distributed by CBS Publishers & Distributors Pvt. Ltd.

Sales & Distribution Rights only for India, Pakistan, Bangladesh, Sri Lanka, Nepal and Bhutan.This book is not to be sold outside these territories.

Contents

List of Abbreviations

AMD	Acid mine drainage
DWR	Department of Water Resources'
DEM	Digital Elevation Model
IEC	Electrotech- nical Commission
FCT	Foundation for the Science and Technology
GIS	Geographic Information System
GPS	Global Positioning System
GMS	Groundwater Modeling System
HRU	Hydrologic response units
HRU	Hydrologic response units
IMS	Integrated management system
IGWMC	International Ground Water Modeling Center
IDW	Inverse Distance Weighting
KPK	Khyber Pakhtunkhwa
MSSs	Man- agement systems standards
MSL	Mean Sea Level
MOSTI	Ministry of Science, Technology and Innovation
NSF	National Science Fellowship
OHS	Occupational health and safety
QA	Quality assurance
QC	Quality control
QMS	Quality Management System
RQD	Rock quality designation
RMSE	Root Mean Square Error
SAFCA	Sacramento Area Flood Control Agency
SRCSD	Sacramento Regional County Sanitation District
SRM	Sacramento Regional Model
SRF	Stress reduction factor
TIN	Triangulated Irregular Network
USDA	United States Department of Agriculture
UTM	Universal Transverse Mercator
WGS	World Geodesic System

List of Contributors

Bizhan Shirdel
Department of Geology, Karaj Branch, Islamic Azad University, Karaj, Iran

Sayed A. Selim
Geology Department, Faculty of Science, Aswan University, Aswan, Egypt

Ali M. Hamdan
Geology Department, Faculty of Science, Aswan University, Aswan, Egypt

Ahmed Abdel Rady
Aswan Water and Sanitation Company, Aswan, Egypt

Saad Merayyan
California State University, Sacramento, CA, USA

Samsor Safi
Sacramento Area Sewer District, Sacramento, CA, USA

Sarva Mangala PRAVEENA
School of Science and Technology, Universiti Malaysia Sabah, Kota Kinabalu, Sabah, Malaysia.

Mohd Harun ABDULLAH
School of Science and Technology, Universiti Malaysia Sabah, Kota Kinabalu, Sabah, Malaysia.

Ahmad Zaharin ARIS
Department of Environmental Sciences, Universiti Putra Malaysia, Selangor, Malaysia

Kawi BIDIN
School of Science and Technology, Universiti Malaysia Sabah, Kota Kinabalu, Sabah, Malaysia.

Jacques Édoukou Djémin
Department of Sciences and Water Technology and Environment Engineering (Laboratory of Remote Sensing and Spatial Analysis Applied to Hydrogeology), Félix Houphouët-Boigny University, Abidjan, Côte d'Ivoire

Jean Kan Kouamé
Department of Sciences and Water Technology and Environment Engineering (Laboratory of Remote Sensing and Spatial Analysis Applied to Hydrogeology), Félix Houphouët-Boigny University, Abidjan, Côte d'Ivoire

Kouakou Serge Deh
Department of Sciences and Water Technology and Environment Engineering (Laboratory of Remote Sensing and Spatial Analysis Applied to Hydrogeology), Félix Houphouët-Boigny University, Abidjan, Côte d'Ivoire

Amani Tawa Abinan
Department of Sciences and Water Technology and Environment Engineering (Laboratory of Remote Sensing and Spatial Analysis Applied to Hydrogeology), Félix Houphouët-Boigny University, Abidjan, Côte d'Ivoire

Jean Patrice Jourda
Department of Sciences and Water Technology and Environment Engineering (Laboratory of Remote Sensing and Spatial Analysis Applied to Hydrogeology), Félix Houphouët-Boigny University, Abidjan, Côte d'Ivoire

Manuel Ferreira Rebelo
CLEGI, Lusíada University, Vila Nova de Famalicão, Portugal

Gilberto Santos
CLEGI, Lusíada University, Vila Nova de Famalicão, Portugal
College of Technology, Polytechnic Institute of Cávado and Ave, Barcelos, Portugal

Rui Silva
CLEGI, Lusíada University, Vila Nova de Famalicão, Portugal
College of Technology, Polytechnic Institute of Cávado and Ave, Barcelos, Portugal

Abdollah Yazdi
Department of Geology, Kahnooj Branch, Islamic Azad University, Kerman, Iran

Mohammad Ali Arian
Department of Geology, North Tehran Branch, Islamic Azad University, Tehran, Iran

Mahmoud M. Rezapour Tabari
Department of Engineering, Shahrekord University, Shahrekord, Iran

Leila Khodapanah
Department of Environmental Sciences, Faculty of Environmental Studies, Universiti Putra Malaysia, Serdang, Malaysia

Wan Nor Azmin Sulaiman
Department of Environmental Sciences, Faculty of Environmental Studies, Universiti Putra Malaysia, Serdang, Malaysia

Hamid Reza Nassery
Faculty of Earth Science, Shahid Beheshti University, Tehran, Iran

Arshad Ashraf
Water Resources Research Institute, National Agricultural Research Center, Islamabad, Pakistan

Luc Descroix
IRD / UJF, Grenoble, France

Ibrahim Bouzou Moussa
UAM University, Niamey, Niger

Pierre Genthon
IRD-HSM, Montpellier, France

Daniel Sighomnou
Niger Basin Authority, Niamey, Niger

Gil Mahé
IRD-HSM, Montpellier, France

Ibrahim Mamadou
University of Zinder, Niger

Jean-Pierre Vandervaere
UJF-LTHE, Grenoble, France

Emmanuèle Gautier
Université Paris 8, France

Oumarou Faran Maiga
UAM University, Niamey, Niger

Jean-Louis Rajot
IRD-BIOEMCO, Créteil, France

Moussa Malam Abdou
IRD / UJF, Grenoble, France

Nadine Dessay
IRD-ESPACE-DEV, Montpellier, France

Aghali Ingatan
UAM University, Niamey, Niger

Ibrahim Noma
UAM University, Niamey, Niger

Kadidiatou Souley Yéro
IRD / UJF, Grenoble, France

Harouna Karambiri
2iE International high School, Ouagadougou, Burkina Faso

Rasmus Fensholt
University of Copenhague, Denmark

Jean Alberge
IRD-LISAH, Montpellier, France

Jean-Claude Olivry
IRD, France

Wei-Zu Gu
Institute for Hydrology and Water Resources, Nanjing Hydraulic Research Institutes, Nanjing, China

Jiu-Fu Liu
Institute for Hydrology and Water Resources, Nanjing Hydraulic Research Institutes, Nanjing, China

Jia-Ju Lu
Institute for Hydrology and Water Resources, Nanjing Hydraulic Research Institutes, Nanjing, China

Jay Frentress
Oregon State University, Corvallis, USA

Jian Liu
Faculty of Geosciences and Environmental Engineering, Southwest Jiaotong University, Chengdu, Sichuan 610031, China

Dan Liu
Faculty of Geosciences and Environmental Engineering, Southwest Jiaotong University, Chengdu, Sichuan 610031, China

Kai Song
Faculty of Geosciences and Environmental Engineering, Southwest Jiaotong University, Chengdu, Sichuan 610031, China

Muhammad Aqeel Ashraf,
Department of Chemistry, University of Malaya, Kuala Lumpur 50603, Malaysia

Mohd. Jamil Maah,
Department of Chemistry, University of Malaya, Kuala Lumpur 50603, Malaysia

Ismail Yusoff
Department of Geology, University of Malaya, Kuala Lumpur 50603, Malaysia

Preface

Environmental geology is an applied science concerned with the practical application of the principles of geology in the solving of environmental problems. Engineering geology is the application of the geological sciences to engineering study for the purpose of assuring that the geological factors regarding the location, design, construction, operation and maintenance of engineering works are recognized and accounted for. The text Environmental, Groundwater, and Engineering Geology focuses on new concepts and recent advances in environmental, groundwater, and engineering geology. Analysis of engineering geology indices in three units of Atamir Formation in Kope Dagh Zone has been focused in first chapter. Second chapter examines the rise in the level of the groundwater in the Quaternary aquifer at Aswan city, Upper Egypt. Third chapter evaluates the feasibility of groundwater banking in the Central Basin. Fourth chapter reviews the types of groundwater solution techniques in terms of advantages and limitations. Fifth chapter presents the assessment of the groundwater intrinsic vulnerability in the Dabou region from the DRASTIC method which requires seven hydrogeological parameters in its application. The aim of sixth chapter is to present and justify a designed methodology to be used by organizations to support the integration of various management systems (MSs). Seventh chapter studies Hormoz Island in terms of geological features and geotourism potentials. The objective of eighth chapter is to develop the hydrogeological framework of the groundwater system in Shariar, Iran and to estimate groundwater balance as a scientific database for future water resources development programs. In ninth chapter, SWAT model developed by United States Department of Agriculture (USDA) has been used to evaluate surface runoff generation, soil erosion and quantify the water balance of a Himalayan watershed in the Northern Pakistan. The purpose of tenth chapter is to provide an overview of hydrological behavior throughout West Africa based on point, local, meso and regional scales observations. Current challenges in experimental watershed hydrology have been revealed in eleventh chapter. Last chapter focuses on evaluating the influence resulting from tunnel excavation on groundwater environment, by means of employing an indicator system.

Chapter 1

ANALYSIS OF ENGINEERING GEOLOGY INDICES IN THREE UNITS OF ATAMIR FORMATION IN KOPE DAGH ZONE

Bizhan Shirdel

Department of Geology, Karaj Branch, Islamic Azad University, Karaj, Iran

ABSTRACT

This is a field report on a comprehensive study of the Atamir Formation from the engineering geology perspective using the related indices. The Atamir Formation of the Cretaceous Period, which has outcropped in the form of thick frequencies of grey-knotted sandstone and black shales, is situated in the Kope Dagh zone. A survey of discontinuities together with bedding was carried out to study slope stability. The layers have a general east-west trend with a gentle slope towards the south. Because of the tectonic and stratigraphic differences, and with the purpose of facilitating surveys related to joint study of the outcrop, the formation in the study region was divided into three units. The lower unit is made of shale, the middle of sandstone, and the upper of marlstone. All three units were studied from the perspective of geomechanical classification, rock mass indices, geological strength, geomechanical indices, and wedge instability analysis under dry and wet conditions, and the results were investigated in the form of various images and figures. The Dips software was used to display the rose diagram and stereographic projection of each unit, the Swedge software to analyze instability of the wedges, and the Roctab software to analyze the geomechanical parameters and present the outputs along with the description of each unit.

INTRODUCTION

Based on field surveys and geological surveys such as outcrop studies, the Atamir Formation was divided into three separate units to study the characteristics of the discontinuities and of the rock mass in each one. This study intended to review results of research carried out on the Atamir Formation in detail and present a field report. Therefore, the units were studied and the results were compared.

RESULTS

The Atamir Formation Unit One K (at) 1

This lithology included black shales. In the study region, there were three dominant discontinuity sets (two joint sets together with bedding) and random discontinuities. Table 1 lists the characteristics of these discontinuities. Table 2 lists geotechnical characteristics of the various discontinuities in the rock masses. Figures 1-3 show the Stereographic characteristics of Figures 4-7 show statistical analysis.

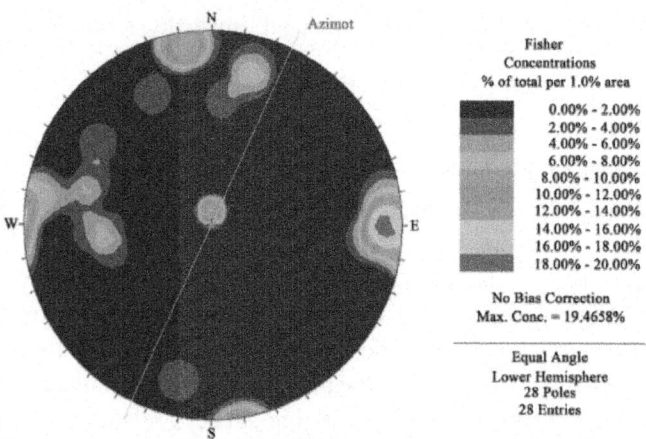

Figure 1: Stereographic image of the discontinuities in unit one of the Atamir Formation K (at) 1.

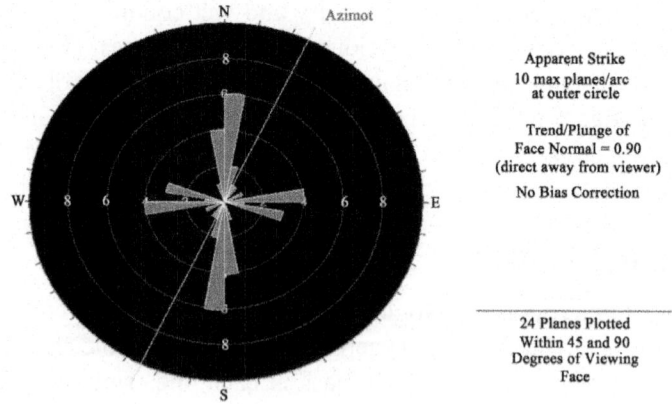

Figure 2: Rose diagram of the discontinuities in unit one of the Atamir Formation K (at) 1.

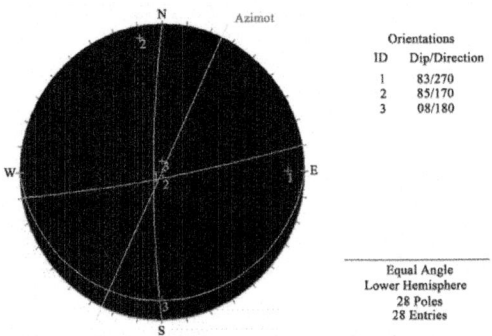

Figure 3: Stereography of discontinuities in unit one of the Atamir Formation K (at) 1.

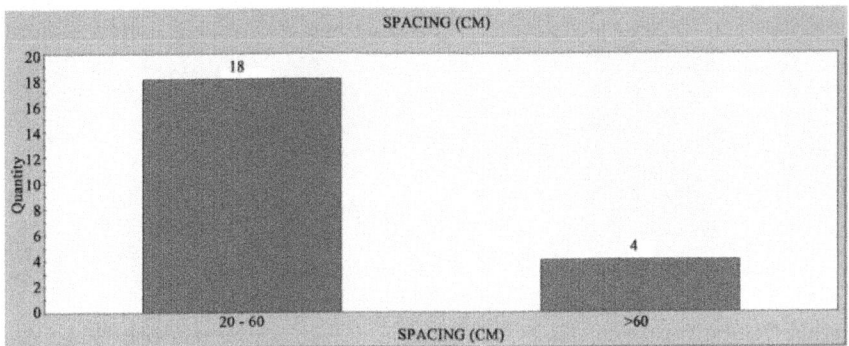

Figure 4: Statistical analysis of discontinuity spacings in unit one of the Atamir Formation K (at) 1.

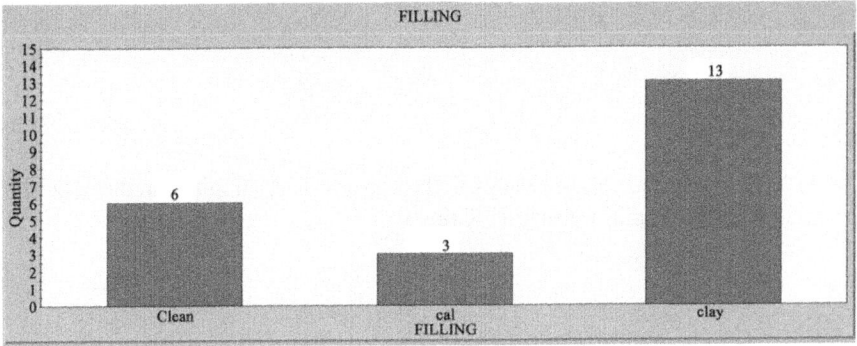

Figure 5: Statistical analysis of discontinuity fillings in unit one of the Atamir Formation K (at) 1.

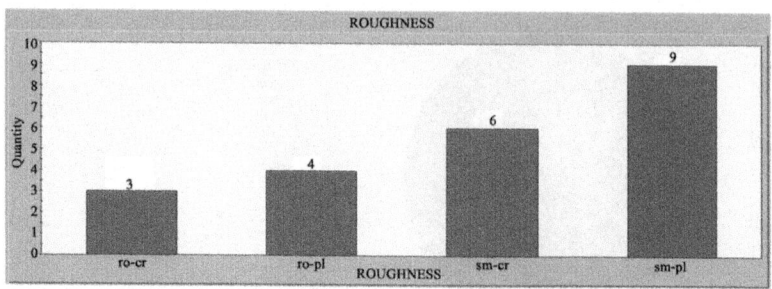

Figure 6: Statistical analysis of discontinuity roughness in unit one of the Atamir Formation K (at) 1.

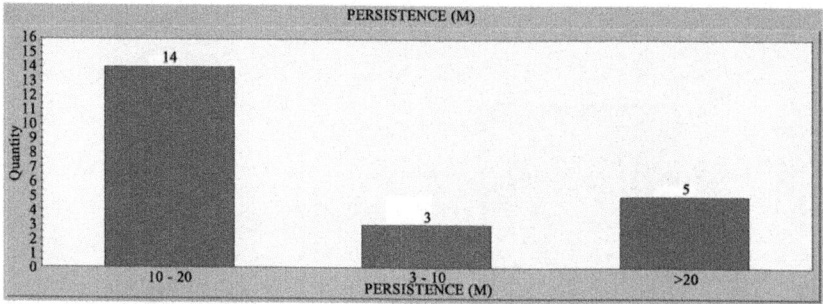

Figure 7: Statistical analysis of discontinuity orientation in unit one of the Atamir Formation K (at) 1.

Table 1: Characteristics of the discontinuities in unit one of the Atamir Formation K (at) 1.

Discontinuity	ID	Dip	Dip Direction
Js1	1	83	270
Js2	2	85	170
Bedding	3	8	180

Table 2; Geotechnical characteristics of the various discontinuities in the rock masses of unit one in the Atamir Formation K (at) 1

Joint Sets	Dip Direction	Joint Frequency	Persistence (m)					Joint Roughness (%)				
			>20%	10% - 20%	3% - 10%	1% - 3%	<1%	Very Rough	Rough	Rough-Smooth	Smooth	Slickenside
Js1	270\83	50	-	80	20	-	-	-	-	20	80	-
Js2	170\85	30	-	70	30	-	-	-	30	70	-	-
Bedding	180\8	20	100	-	-	-	-	-	-	100	-	-

Joint Alteration	Joint Water	Aperture (mm)					Joint filling					Joint spacing (m-mm)					
		>5 mm %	1 - 5 mm %	0.1 - 1 mm %	<0.1 mm %	Close	Soft filling mm 5 < %	Soft filling mm 5 > %	Hard filling mm 5 > %	Hard filling mm 5 < %	Clean %	<60 mm %	60 - 200 mm %	200 - 600 Mm %	0.6 - 2 m %	>2 m %
		-	100	-	-	-	-	100	-	-	-	-	-	100	-	-
W2	Dry	40	60	-	-	-	-	-	40	-	60	-	-	50	50	-
		-	100	-	60	-	100	-	-	-	-	-	-	-	-	100

Rock masses were classified based on the methods proposed by Bieniawski (RMR), Barton (Q), and Hoek (GSI), all of which are discussed below.

Geomechanical Classification (RMR)

- Compressive strength of intact rock
    - The value of this parameter was determined based on the strength characteristic of outcropped masses and on performing tests.
- Rock quality designation
    - The rating of 8 may be given depending on the state of the discontinuity system.
- Joint spacing
    - The rating for this parameter was the least value of spacing. In the study region, there were three sets of discontinuities (two sets of dominant discontinuities together with bedding), and the minimum discontinuity spacing was less than 10 centimeters. Considering the geomechanical classification parameters, the rating of 8 - 10 was given to this parameter.
- Condition of discontinuities
    - The joints in the rock masses in unit one of the Atamir Formation, K (at) 1, were 10 - 20 meters long, clay, calcite, and few traces of iron oxide were the main filling, the joints were smooth to a little rough and, with respect to euhidrality, the rock masses were planar and moderately weathered. Considering the existing parameters, the rating given to this parameter was 10.
- Groundwater
    - Presence of groundwater in a region is undoubtedly very important, especially if permeability is high. Considering the conditions of the rock masses in the region, the seepage state was determined and given the rating of 4.
- Discontinuity orientations

This parameter is determined by the mode of transgression of the structure towards the jointed rock mass, by the dips of the layers (and of the discontinuities), and by the orientation of the structure (which has been used by Feenstra & Wickham in 1975 [1] and suggested by Bieniawski 1973) [2] .

The presented parameters 2 - 1 to 2 - 5 are used for determining the value of RMR that, for the rock masses in unit one of the Atamir Formation, K (at) 1, was 34 - 36. Based on this, these rock masses were placed in the poor class; pertaining data are shown in Table 3.

Qualitative Index of the Rock Mass (Q)

The index Q for the rock masses in unit one of the Atamir Formation, K9 (at) 1, was estimated based on the following 6 parameters.

- Rock quality designation (RQD)

  Considering the condition of the rock masses, the value of 43 - 48 was given to the RQD parameter of unit one in the Atamir Formation, K (at) 1.

Table 3. Geomechanical classification of the rock masses in unit one of the Atamir Formation K (at) 1

Parameter	Value	Rating
UCS (Mpa)	25 - 50	4
RQD	25 - 50	8
Spacing (mm)	60 - 200	8 - 10
Condition of discontinuities		10
Groundwater	Flow	4
RMR	34 - 35	
Class no.	Poor rock	
Cohesion of the rock mass (KPa)	115	
Friction angle of the rock mass (deg)	15 - 25	

- Number of joint sets (J_n)

  Based on the data obtained from the geological field surveys, this parameter received the value of 12 for unit one of the Atamir Formation, K (at) 1.

- Discontinuity surface roughness (J_r)

  This parameter shows the degree of roughness of the discontinuities and, based on Barton's theory, the weakest set of critical discontinuities or discontinuities filled with clay materials in the related zone must be determined. Considering the conditions, the value of one was given to this parameter for unit one of the Atamir Formation, K (at) 1.

- Joint alteration number (J_a)

  Considering the joint filling and the apparent conditions of discontinuity surfaces, this parameter received the value of 4 for unit one of the Atamir Formation, K (at) 1.

- Joint water reduction factor (J_w)

This parameter measures water pressure, which inversely affects shear strength of the discontinuity. Based on data collected in the study region, the value of 0.66 was considered for this parameter.

- Stress reduction factor (SRF)

   This parameter is determined based on the characteristics of the study region including the existing stresses, the tectonic conditions, the shear zones, etc. Since the necessary tests were not performed, the estimated value of one was used for this factor.

Considering the ratings of the six parameters, and using the presented relationship, the expected value determined for the Q index was 0.592 - 0.66 for the rock masses in unit one of the Atamir Formation, K (at) 1, which places them in the class of very poor rocks. Data are shown in Table 4.

Geological Strength Index (GSI) of the Rock Masses

Using the following three methods, the value of the geological strength index for the rock masses in unit one was determined.

- Geological strength index of the rock mass (GSI) (Figure 8)

   The GSI system is the only classification system that is directly related to engineering parameters such as Mohr-Coulomb and Hoek-Brown parameters. Conventional methods for determining strength parameters and rock mass modulus in Iran and other countries include plate-loading tests and block shear in situ. However, these tests can be performed only if exploration drilling has been carried out and, in addition to that, they are very costly. Therefore, many studies have been conducted and numerous attempts made for developing and introducing methods to indirectly determine engineering parameters of rock masses, one of which is the GIS system [3] . In the following sections, the GSI of rock masses in unit one will be discussed.

- Determining GSI based on field observations

   Based on field studies and observations, the value of geological strength index of rock masses in unit one of the Atamir Formation, K (at) 1 was 35 - 45.

Table 4: The Q index of the rock masses in unit one of the Atamir Formation, K (at) 1

Parameter	Description	Value
RQD	43 - 48	43 - 48
J_n	Three joint sets with random	12
J_r	Smooth or rough to **Craw** or Planar	1
J_a	Softening or low friction clay mineral coatings	4
J_w	Inflow	0.66
SRF	Medium stress, favorable stress condition	1
Q	0.592 - 0.66	
Class No.	Very poor rock	

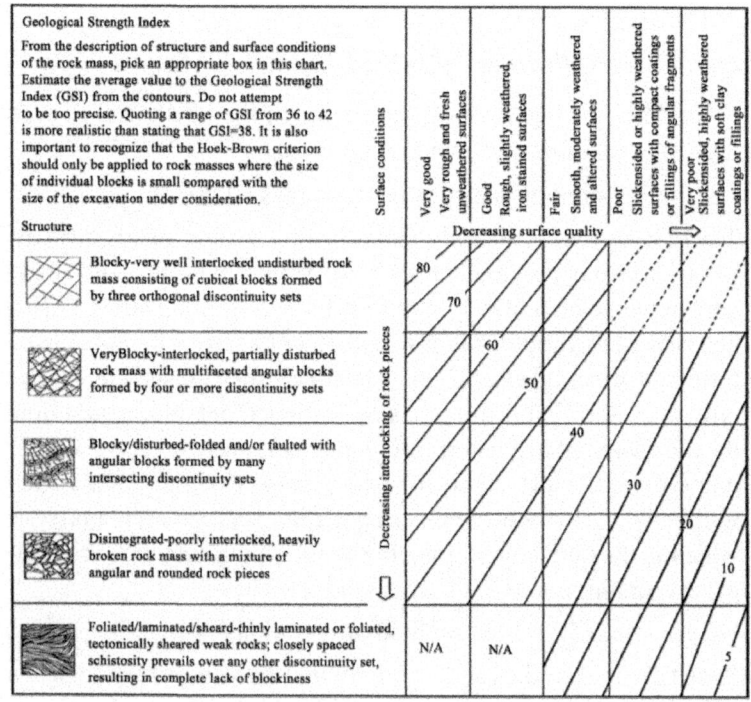

Figure 8: The GSI range of the rock masses in unit one of the Atamir Formation K (at) 1 in the GSI classification diagram.

Determination of Geomechanical Parameters of the Rock Masses in Unit One of the Atamir Formation K (at) 1

To determine these parameters, the GSI was considered 35 - 45. Since cohesive strength and angle of internal friction are among the input parameters in most methods of analysis, the Mohr-Coulomb and Hoek-Brown criteria had to be fitted to determine these two parameters. Results are listed in Table 5 and shown in Figures 9-11.

Analysis of Wedge Instability in Unit One of the Atamir Formation, K (at) 1, under Dry and Wet Conditions

Analysis of wedge instability was performed using SWEDGE. This software, which was developed by the ROCSCIENCE Company, analyzes wedge stability in rock slopes, degree of cohesion, geomechanical characteristics, etc., studies stability from various aspects, analyzes areas with rock fall potential using the StereoNet network, follows discontinuities and determines and analyzes intersection points of wedges, and plots the StereoNet using input data. Figure 12 shows stereographic images of the joint sets. Applying the primary parameters, the characteristics of the largest wedges that can be formed in the walls are determined. According to the performed analysis, wedge slide will happen along the interface between two discontinuities, and the factor of safety (FS) will be 0.07743. Figures 13-18 show other views.

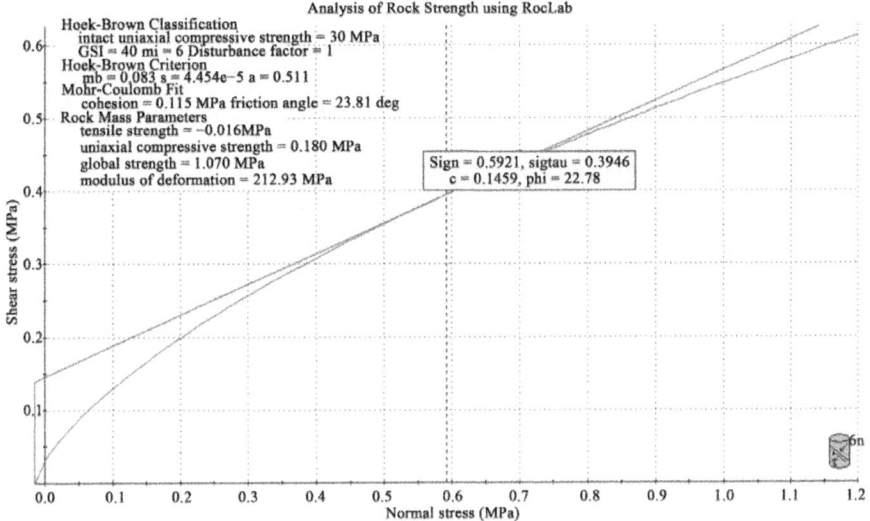

Figure 9: Fitting the data on the rock masses in unit one of the Atamir Formation, K (at) 2, to the Hoek-Brown and Mohr- Coulomb criteria based on GSI = 40 and normal shear stress.

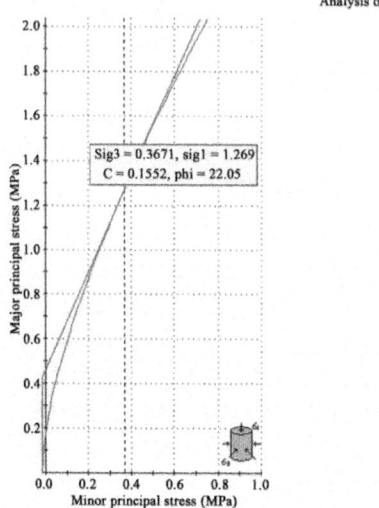

Figure 10: Fitting the data on the rock masses in unit one of the Atamir Formation, K (at) 2, to the Hoek-Brown and Mohr- Coulomb criteria based on GSI = 40 and major principal stress.

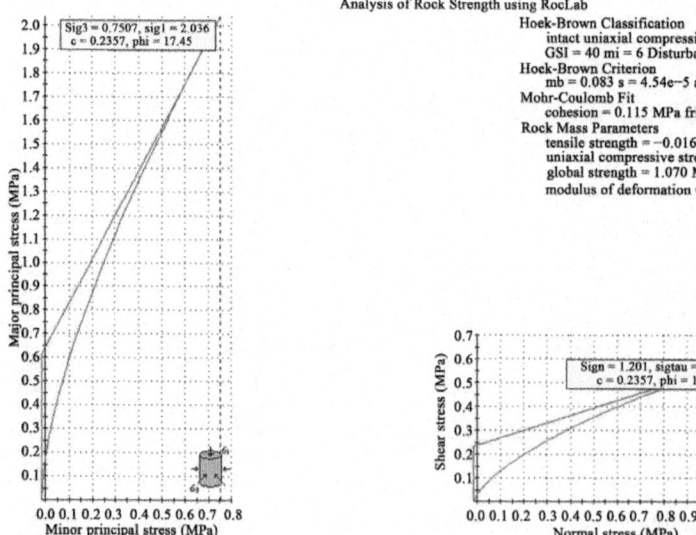

Figure 11: Analysis of output data for determining geomechanical parameters of the rock masses in unit one based on GSI = 40.

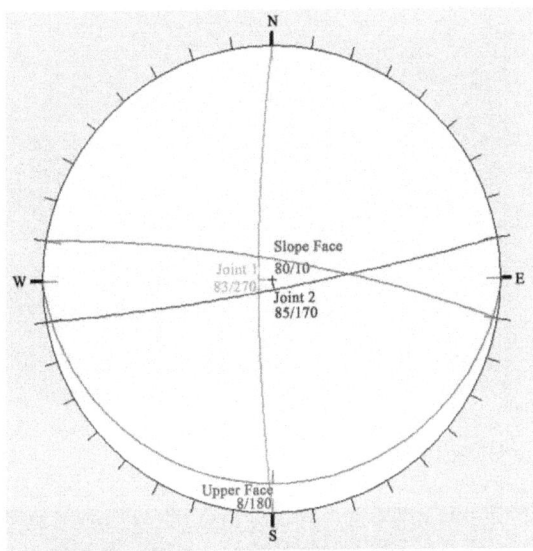

Figure 12: Stereographic images of joint sets.

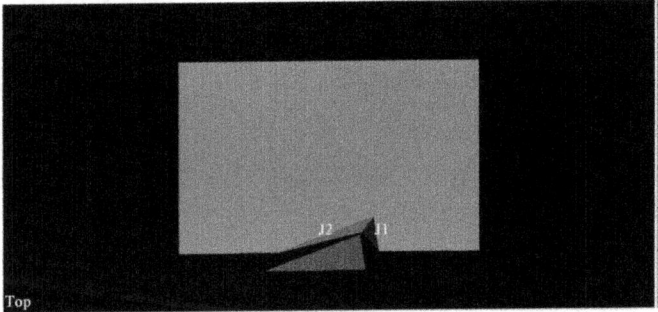

Figure 13: Top view of wedge slide analysis.

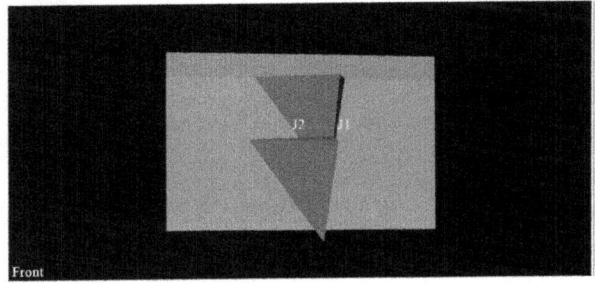

Figure 14: Front view of wedge slide analysis.

Figure 15: Wedge slide: front view.

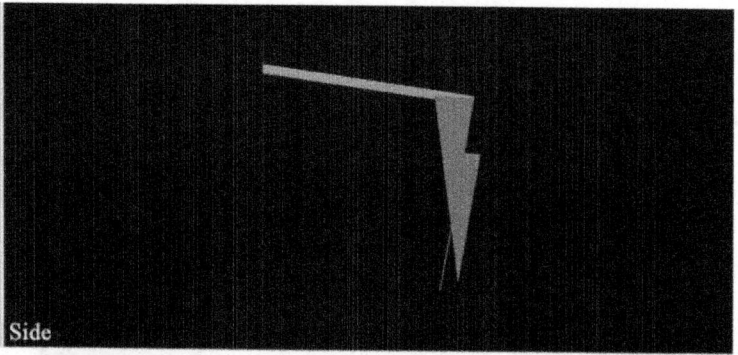

Figure 16: Side view of wedge slide analysis.

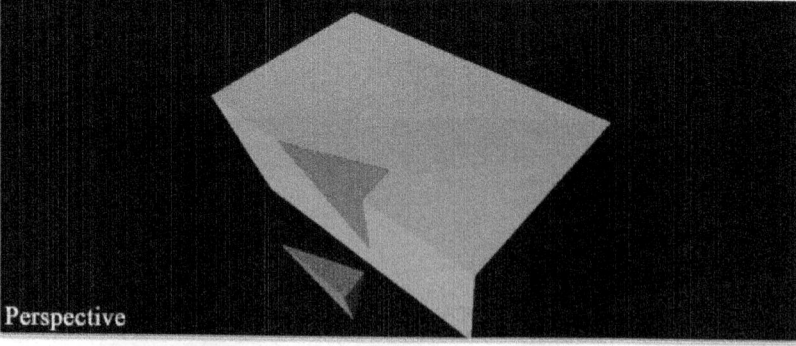

Figure 17: Perspective view of wedge slide analysis.

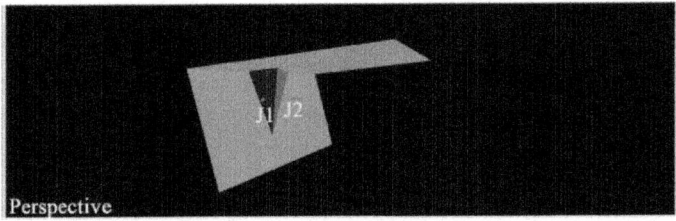

Figure 18: A view of wedge slide analysis obtained by applying directions of water penetration.

Table 5: Geotechnical parameters of the rock masses in unit one of the Atamir Formation K (at) 1

	Parameter		Min	Max	Ave
	GSI		35	45	40
	Intact uniaxial compressive strength (Mpa)		10	50	30
	Mi		3	9	6
Material constants		m_b	0.0235	0.923	0.083
		s	2.65	6.73	4.54
		A	0.515	0.502	0.511
	Cohesion (Mpa)		0.089	0.431	0.115
	Friction angle (Deg)		12.76	62.76	23.81
	Tensile strength		−0.010	−0.05	−0.016
Rock mass parameter (MPa)	Uniaxial compressive strength		0.09	0.98	0.180
	Global strength		0.9	2.1	1.070
	Modulus of deformation		180.97	560.98	212.93

Analysis of Wedge Instability under Wet Conditions in Unit One of the Atamir Formation, K (at) 1

Although water is not directly involved in mass (slope) movement, it is an important factor because:

- Increased water content of slopes caused by rain and snowmelt makes them heavy. Water can penetrate into pores and fractures and replace air and, since water is heavier than air, weights of rocks increase. Because weight is the same as force, and force applies stress on surfaces, increased stress can cause instability.

- Water can alter slope angle (angle of the slope at which the slope is stable).

- Water can be absorbed or repelled by soil mineral substances. Water absorption causes the electrical poles of water molecules to stick to the surfaces of mineral materials and penetrate into them. Therefore, increased water content makes rocks heavier, leading to reduced rock strength. In general, wet clay soils have less strength compared to dry clay soils. Therefore, water absorption reduces strength and, since clay is the material that fills the spaces between the joints in this unit, increased water content gains more importance.

- Groundwater is present almost everywhere under the surface of the ground. This water fills the empty spaces between rock particles and even the cracks within rocks. Groundwater level changes due to rainfall and rises in wet seasons leading to greater water penetration. In dry seasons, groundwater level falls causing less water penetration. These changes in groundwater levels can be an important factor in slope stability.

If water penetrates through joint surfaces, the factor of safety will become zero (FS = 0) (Figure 19).

Unit Two of the Atamir Formation, K (at) 2 (from 300 + 3 to 500 + 5 Meters)

This lithology includes dark grey sandstones, there are three dominant discontinuity sets in the study region (two joint sets together with bedding), and the discontinuities are random. Table 6and Table 7 shows the characteristics of the discontinuities. It is shown in Figures 20-22.

Figure 19: Water penetration into joint sets in unit one of the Atamir Formation.

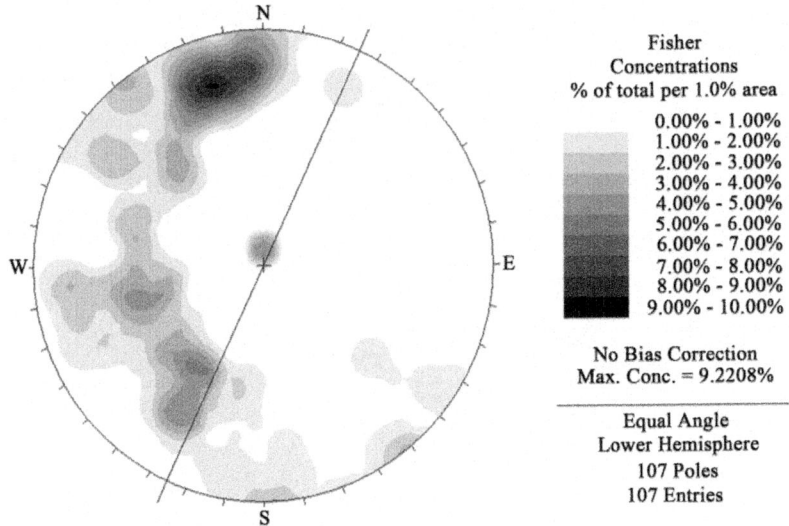

Figure 20: Stereographic image of the discontinuities in unit two of the Atamir Formation K (at) 2.

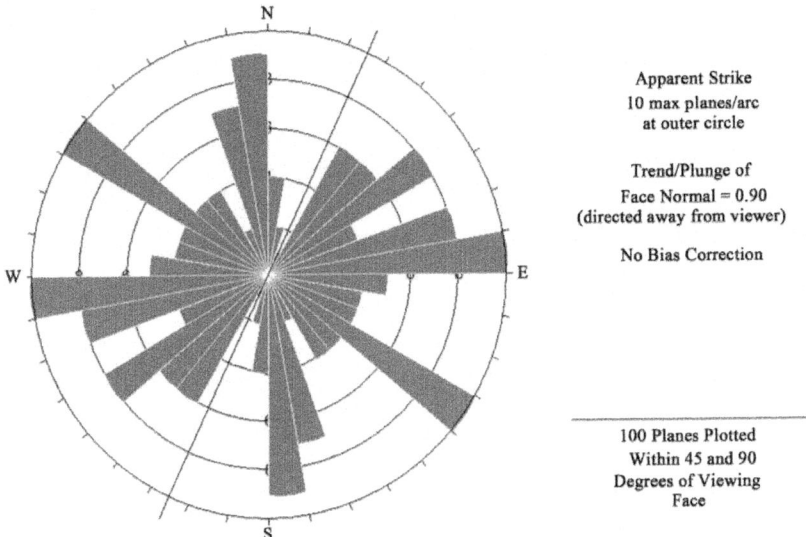

Figure 21: Rose diagram of the discontinuities in unit two of the Atamir Formation K (at) 2.

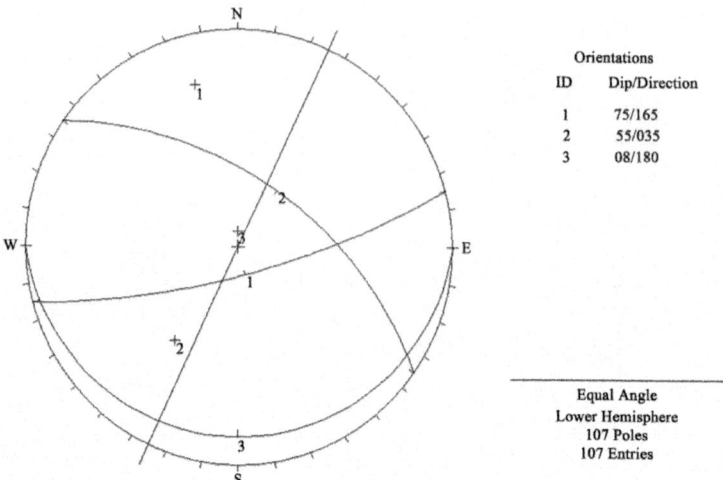

Orientations	
ID	Dip/Direction
1	75/165
2	55/035
3	08/180

Equal Angle
Lower Hemisphere
107 Poles
107 Entries

Figure 22: Stereographic image of the discontinuities in unit two of the Atamir Formation K (at) 2.

Table 6: Characteristics of the discontinuities in unit two of the Atamir Formation K (at) 2

Discontinuity	ID	Dip	Dip direction
Js1	1	75	165
Js2	2	55	035
Bedding	3	8	180

Table 7: Geotechnical characteristics of the various discontinuities in the rock masses of unit two in the Atamir Formation K (at) 2

Joint sets	Dip Dip direction		Persistence (m)						Joint roughness (%)			
			>20%	10% - 20%	3% - 10%	1% - 3%	<1%	Very rough	Rough	Rough-Smooth	Smooth	Slickenside
Js 1	165\75	60	80	20	-	-	-	-	80	20	-	-
Js 2	035\55	20	30	70	-	-	-	-	30	70	-	-
Bedding	180\8	20	100	-	-	-	-	-	100	100	-	-

Joint alteration	Joint water	>5 mm %	Aperture (mm)					Joint filling				Joint spacing (m-mm)				
			1 - 5 mm %	0.1 - 1 mm %	<0.1 mm %	Close	Soft Filling mm 5 < %	Soft Filling mm 5 > %	Hard Filling mm 5 > %	Hard Filling mm5< %	Clean %	<60 mm %	60 - 200 mm %	200 - 600 mm %	0.6 - 2 m %	>2 m %
		90	10	-	-	-	-	50	-	50	-	-	-	100	-	-
W1	Dry	60	40	-	-	-	-	-	70	-	30	-	-	70	30	-
		-	100	-	-	-	70	30	-	-	-	-	-	-	-	100

Figures 23-26 show statistical analysis for unit 2.

Tables 8-17 show information of data for the Atamir Formation K (at) 2 like the order of previous zone.

The GSI range of the rock masses in unit two of the Atamir Formation is shown in Figure 27.

Fitting data related to the rock masses in unit two of the Atamir Formation is shown in Figures 28-30.

Stereographic images of joint sets are shown in Figure 31. Wedge slide analysis is also shown in Figures 32-36.

Unit Three of the Atamir Formation, K (at) 3 (from 500 + 3 to 540 + 3)

This lithology consists of dark grey sandstones, and there are four dominant discontinuity classes (three joint classes together with bedding) and random discontinuities.

Stereography of the discontinuities is shown in Figure 37-39.

Statistical analysis of discontinuity from Figures 40-43.

Range of GSI in the rock masses of unit three in the Atamir Formation is shown in Figure 44.

Fitting the data on rock masses in unit three is shown in Figures 45-47.

Stereographic images of the joint sets for unit 3 is shown in Figure 48.

Views of wedge slide analysis are shown in Figures 49-53.

CONCLUSIONS

All three units had very poor to poor rocks that required rapid stabilization at most 10 hours after blasting operations. The following geomechanical and geotechnical characteristics were observed:

- Layered sedimentary rocks sloping towards the outer domain with slopes close or equal to dip slope.
- Joints and foliations that form weak extended surfaces and intersect the domain surface.
- Intersecting joints that cause wedge slide failure.
- Presence of soft, layered, strongly jointed shales having weak layers.
- Saturation or wetting of joint surfaces having clay filling by water from rainfall, snowmelt, changes in groundwater levels, and reduced frictional force (considering the failure mechanism, and ground surveys,

the effects of last spring's rain on the occurrence of this phenomenon cannot be denied).

- Wet and semi-humid climates and successive snowfalls and, in addition, drought and wet periods.

- Formations such as marl and shale ones, or intermittent marl and shale formations, are susceptible to mass movements, and the more unsaturated clay fills the joints, the more susceptible the formation will be to mass movement.

- The tectonic condition (topography) of the dip slope (slopes exceeding 30%), and slope and orientation of the faults, layers, and fractures.

- Failure to start stabilization in time according to the RMR table.

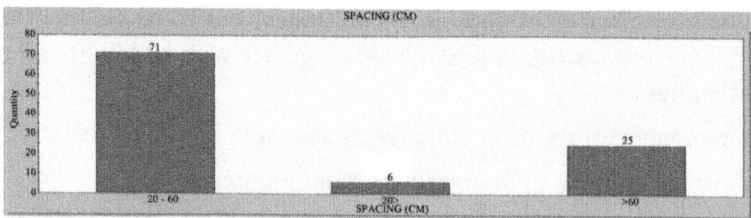

Figure 23. Statistical analysis of discontinuity spacing in unit two of the Atamir Formation K (at) 2.

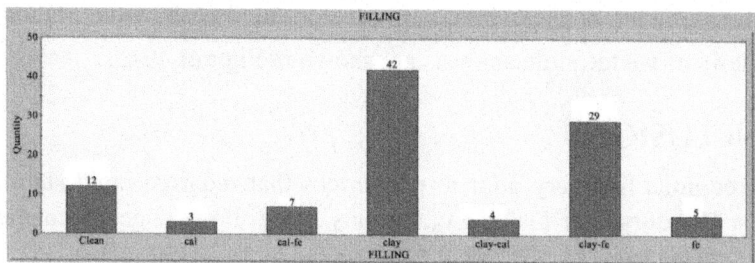

Figure 24. Statistical analysis of discontinuity filling in unit two of the Atamir Formation K (at) 2.

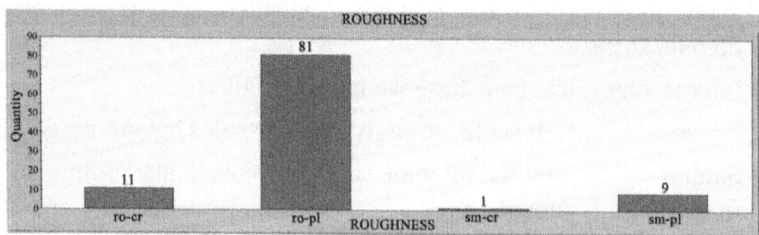

Figure 25. Statistical analysis of discontinuity roughness in unit two of the Atamir Formation K (at) 2.

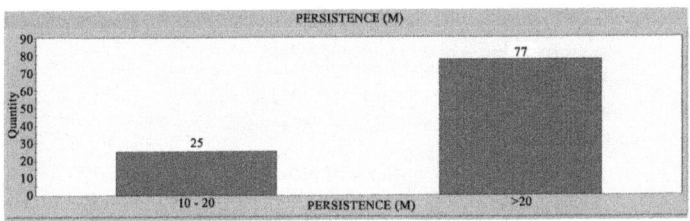

Figure 26. Statistical analysis of discontinuity orientations in unit two of the Atamir Formation K (at) 2.

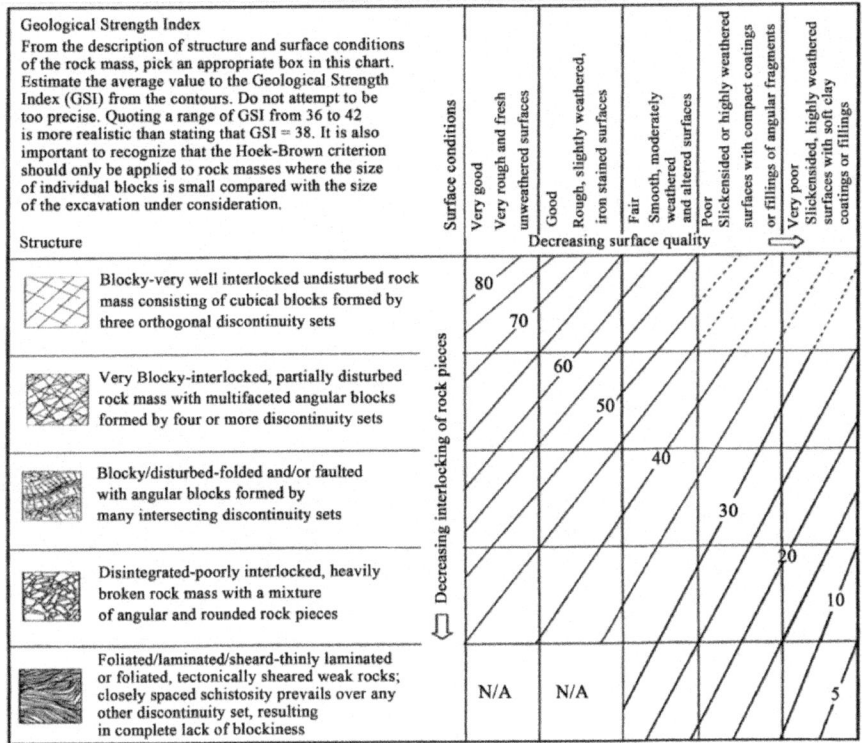

Figure 27. The GSI range of the rock masses in unit two of the Atamir Formation K (at) 2 in the GSI classification diagram.

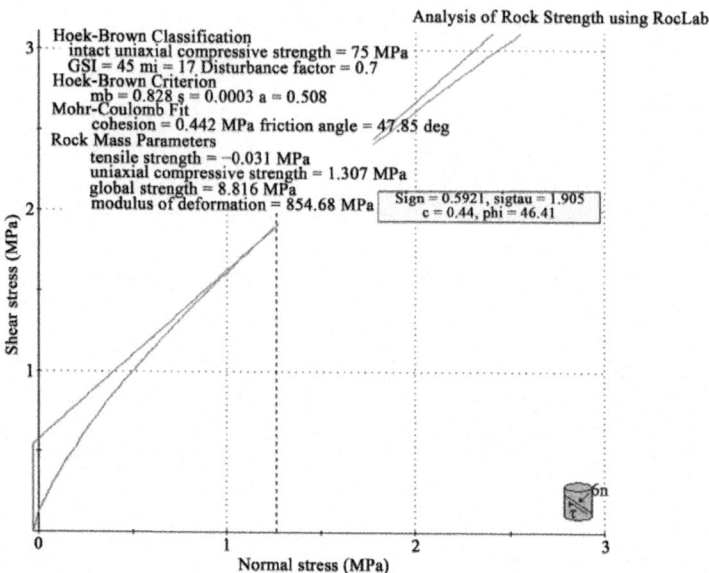

Figure 28. Fitting data related to the rock masses in unit two of the Atamir Formation, K (at) 2, to Hoek-Brown and Mohr- Coulomb criteria.

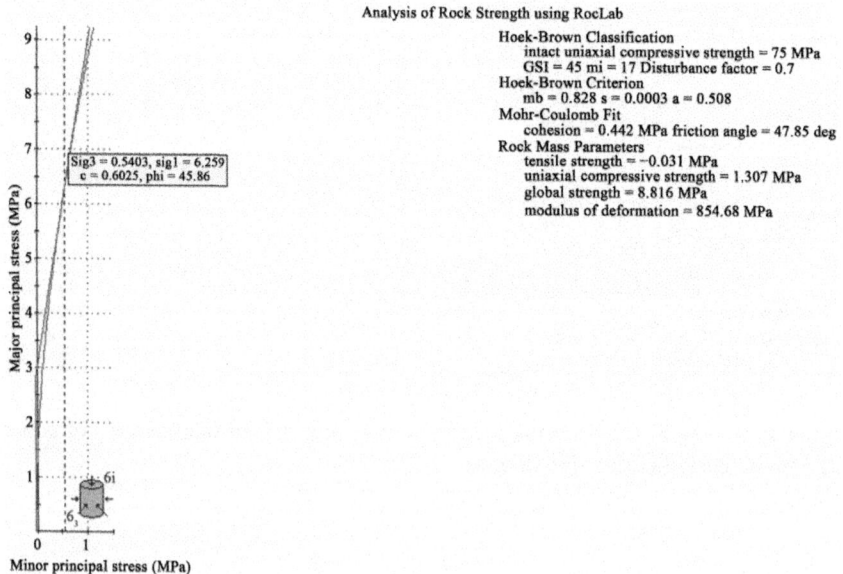

Figure 29. Fitting the data related to the rock masses of unit two to the Hoek-Brown and Mohr-Coulomb criteria based on GSI = 45 and on major principal stress.

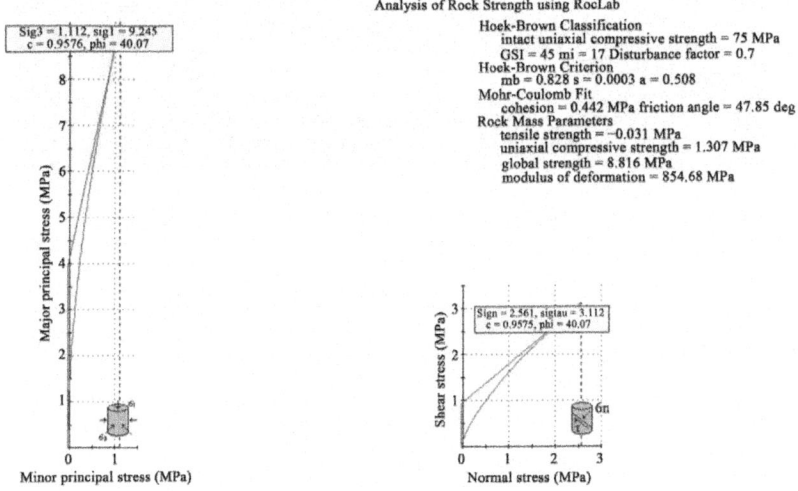

Figure 30. Analysis of output data determining geomechanical parameters of the rock masses in unit two based on GSI = 45.

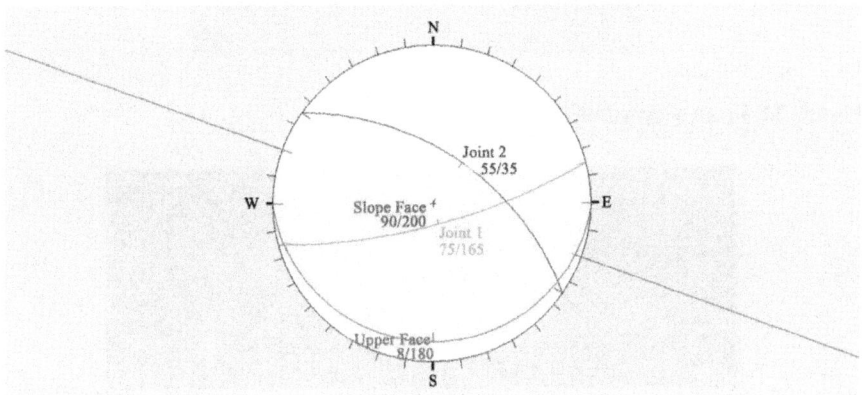

Figure 31. Stereographic images of joint sets.

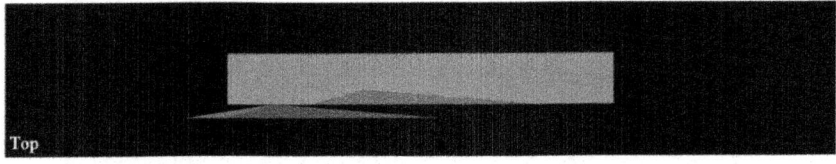

Figure 32. Top view of wedge slide analysis.

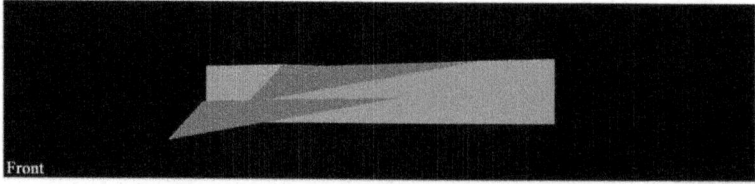

Figure 33. Front view of wedge slide analysis.

Figure 34. Front view of wedge slide.

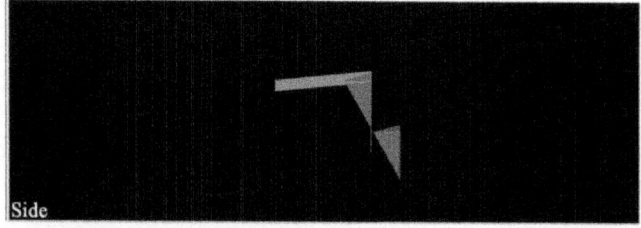

Figure 35. Side view of wedge slide analysis.

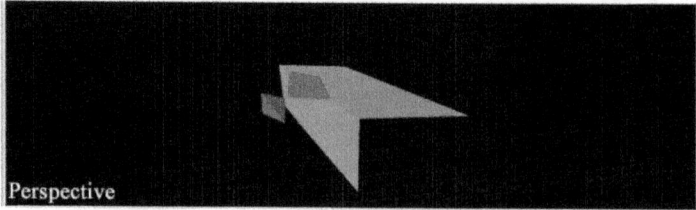

Figure 36. Perspective view of wedge slide analysis.

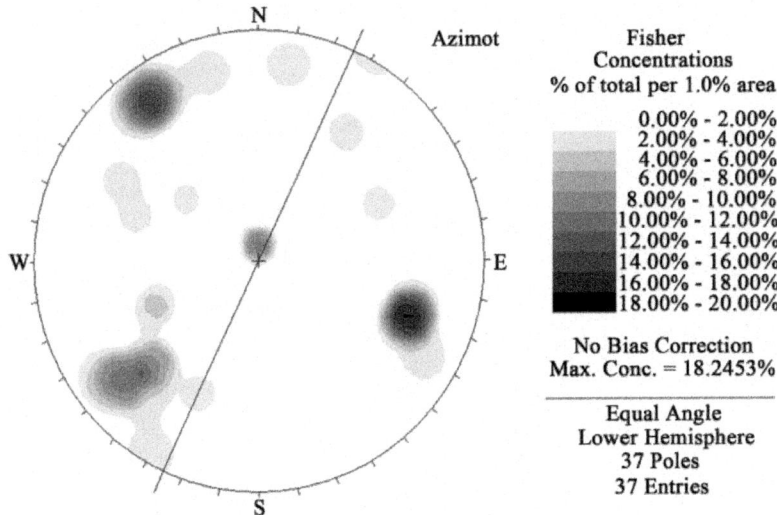

Figure 37. Stereographic image of the discontinuities in unit three of the Atamir Formation K (at) 3.

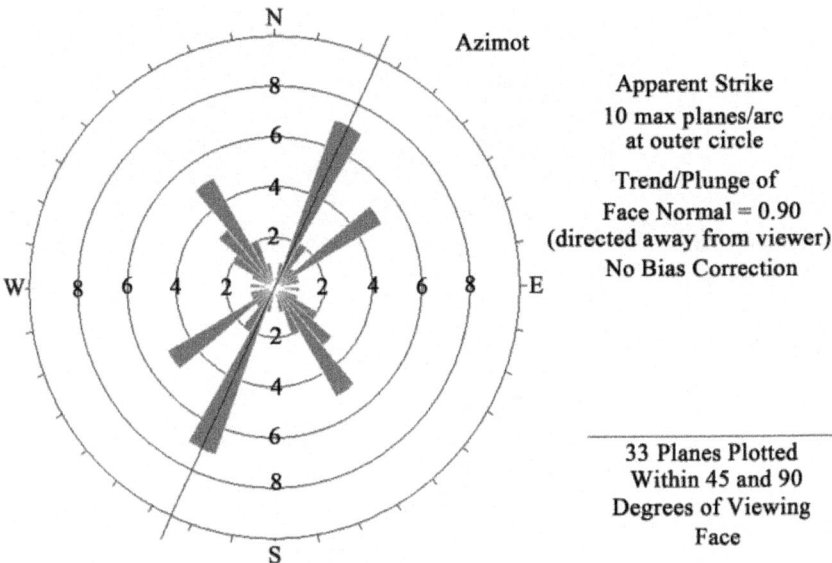

Figure 38. Rose diagram of the discontinuities in unit three of the Atamir Formation K (at) 3.

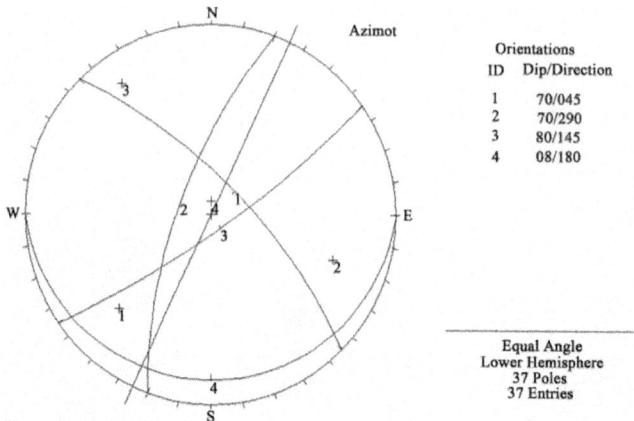

Figure 39. Stereography of the discontinuities in unit three of the Atamir Formation K (at) 3.

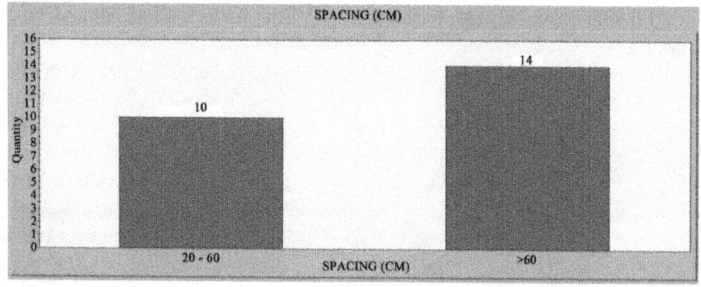

Figure 40. Statistical analysis of discontinuity spacings in unit three of the Atamir Formation K (at) 3.

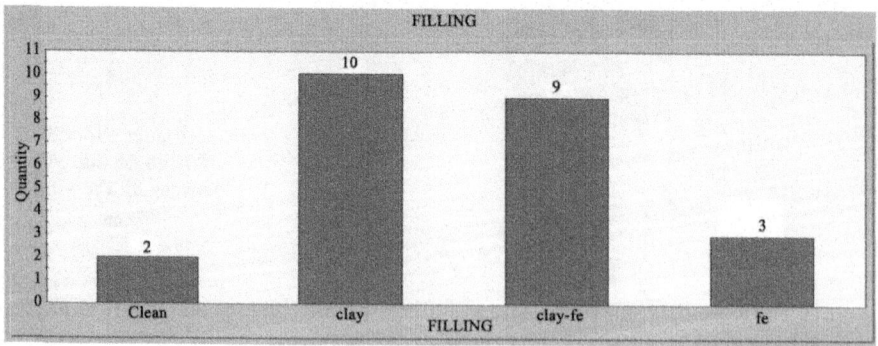

Figure 41. Statistical analysis of discontinuity filling in unit three of the Atamir Formation K (at) 3.

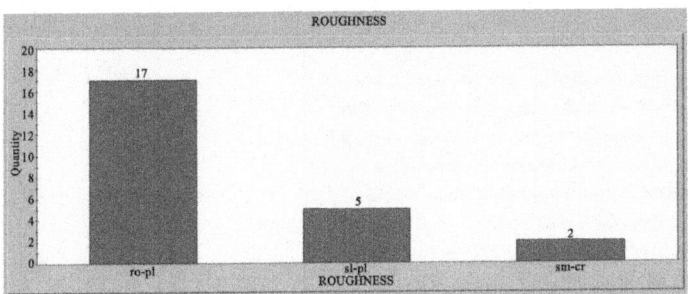

Figure 42. Statistical analysis of discontinuity roughness in unit three of the Atamir Formation K (at) 3.

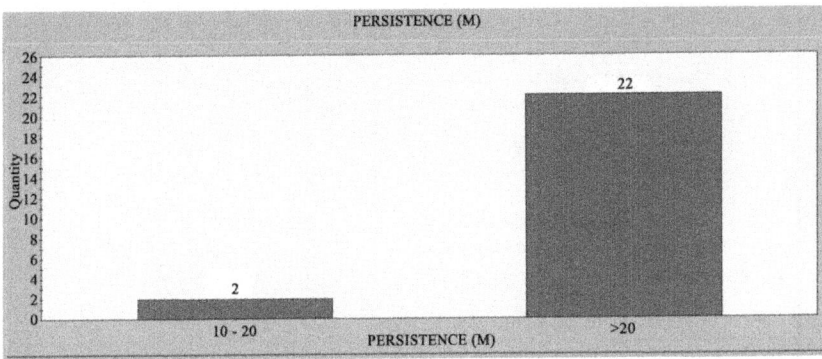

Figure 43. Statistical analysis of discontinuity orientation in unit three of the Atamir Formation K (at) 3.

Table 8. Geomechanical classification of the rock masses in unit two of the Atamir Formation K (at) 2

Parameter	Value	Rating
UCS (MPa)	40 - 60	4 - 6
RQD	25 - 50	8
Spacing (MM)	60 - 200	8 - 10
Condition of discontinuities	-	10
Groundwater	Dam	7
RMR	37 - 41	
Class No.	Dam	
Cohesion of the rock mass (KPa)	442	
Friction angle of the rock mass (deg)	15 - 25	

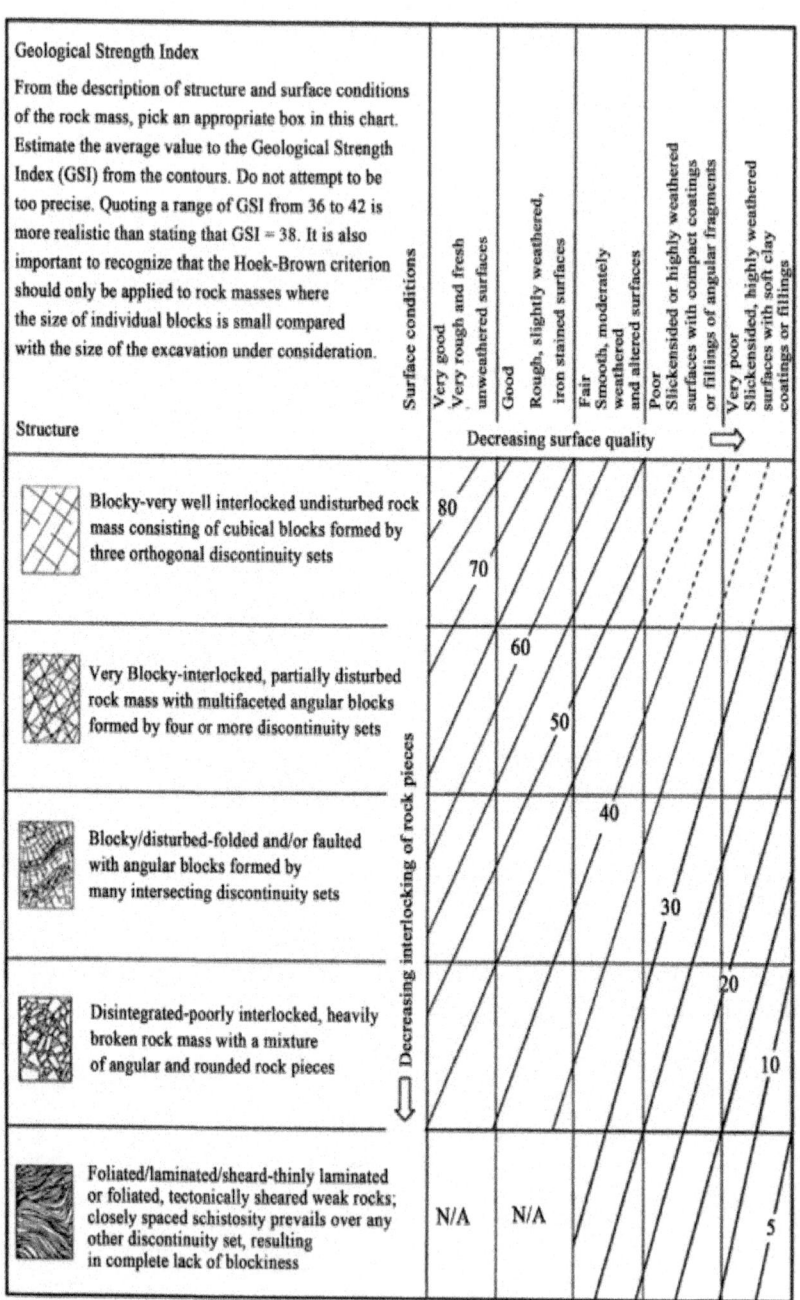

Figure 44. Range of GSI in the rock masses of unit three in the Atamir Formation K (at) 3.

Table 9. The Q index of the rock masses in unit two of the Atamir Formation K (at) 2

Parameter	Description	Value
RQD	40 - 50	40 - 45
J_n	Three joint sets with random to four joint sets	12
J_r	Smooth to planar	1
J_a	Softening or low friction clay mineral coatings	4
J_w	Dry to medium inflow of pressure	1 - 0.66
SRF	Medium stress, favorable stress condition	1
Q	0.62 - 0.83	
Class No.	Very poor rock	

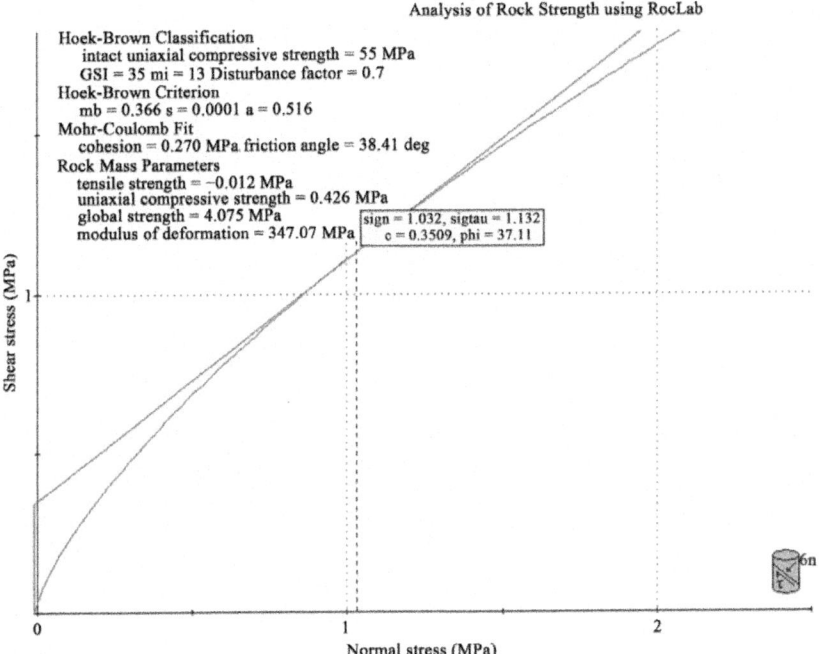

Analysis of Rock Strength using RocLab

Hoek-Brown Classification
   intact uniaxial compressive strength = 55 MPa
   GSI = 35 mi = 13 Disturbance factor = 0.7
Hoek-Brown Criterion
   mb = 0.366 s = 0.0001 a = 0.516
Mohr-Coulomb Fit
   cohesion = 0.270 MPa friction angle = 38.41 deg
Rock Mass Parameters
   tensile strength = −0.012 MPa
   uniaxial compressive strength = 0.426 MPa
   global strength = 4.075 MPa
   modulus of deformation = 347.07 MPa

sign = 1.032, sigtau = 1.132
c = 0.3509, phi = 37.11

Figure 45. Fitting the data on rock masses in unit three to the Hoek-Brown and Mohr-Coulomb criteria, based on GSI = 35 and on normal shear stress.

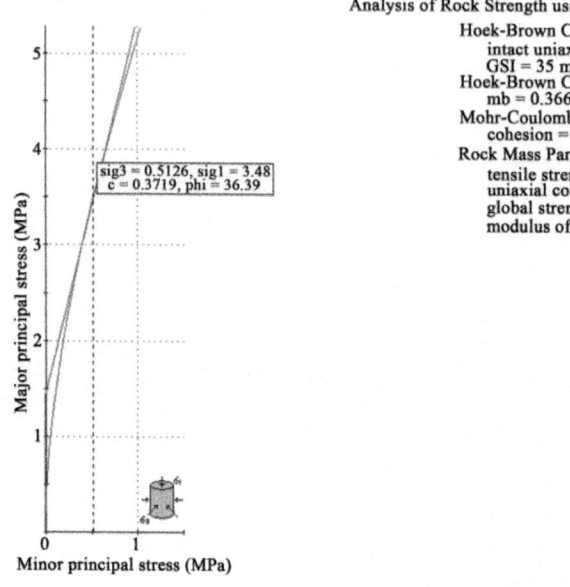

Figure 46. Fitting the data on rock masses in unit three to the Hoek-Brown and Mohr-Coulomb criteria, based on GSI = 35 and on major principal stress.

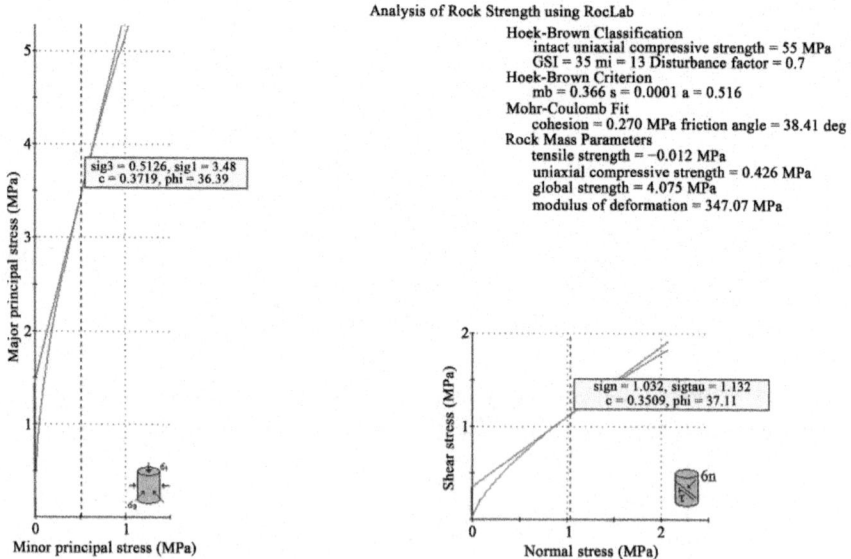

Figure 47. Analysis of output data determining the geomechanical parameters of rock masses in unit three, based on GSI = 35.

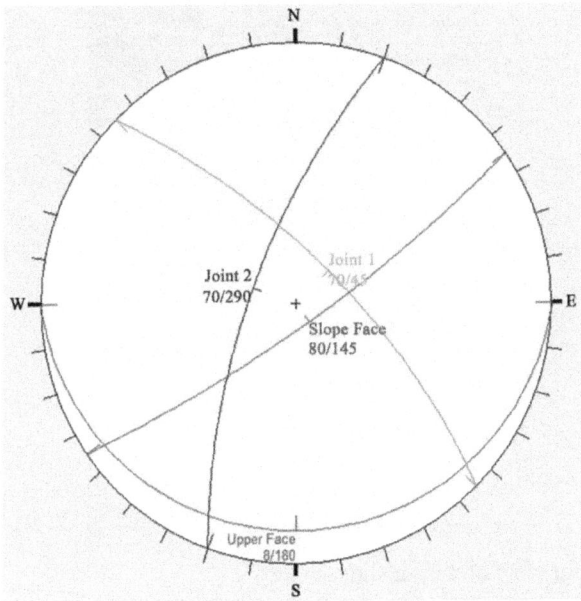

Figure 48. Stereographic images of the joint sets.

Figure 49. Top view of wedge slide analysis.

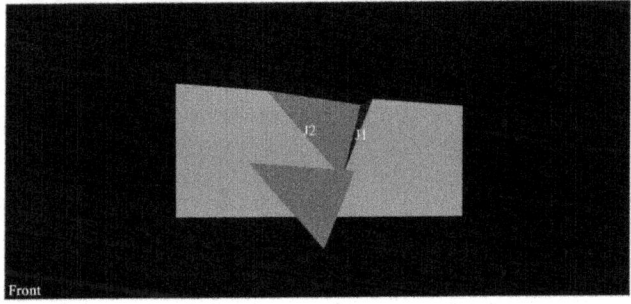

Figure 50. Front view of wedge slide analysis.

Figure 51. Front view of wedge slide.

Figure 52. Side view of wedge slide analysis.

Figure 53. Perspective view of wedge slide analysis.

Table 10. Geomechanical parameters of the rock masses in unit two of the Atamir Formation K (at) 2

Parameter		Min	Max	Ave
GSI		40	50	45
Intact uniaxial compressive strength (Mpa)		50	100	75
Mi		11	23	17
Material constants	m_b	0.154	1.754	0.828
	S	0.0001	0.0014	0.0003
	A	0.478	0.612	0.508
Cohesion (Mpa)		0.210	0.501	0.442
Friction angle (deg)		62.87	90.63	47.85
Rock mass parameter (Mpa)	Tensile strength	−0.019	−0.08	−0.031
	Uniaxial compressive strength	0.096	2.68	1.307
	Global strength	6.98	10.67	8.816
	Modulus of deformation	600.75	1000.5	854.68

Table 11. Characteristics of the discontinuities in unit three of the Atamir Formation K (at) 3

Discontinuity	ID	Dip	Dip direction
Js1	1	70	45
Js2	2	70	290
Js3	3	80	145
bedding	4	8	180

Table 12. Geotechnical characteristics of various discontinuities in rock masses of unit three in the Atamir Formation K (at) 3

Joint sets	Dip Dip direction		Persistence (m)					Joint roughness (%)				
			>20%	10% - 20%	3% - 10%	1% - 3%	<1%	Very rough	Rough	Rough-Smooth	Smooth	Slickenside
Js1	45\70	50	100	-	-	-	-	100	-	-	-	-
Hs2	290\70	25	70	30	-	-	-	-	20	80	-	-
Js3	145\80	10	-	100	-	-	-	-	-	100	-	-
Bedding	180\8	15	100	-	-	-	-	100	-	-	-	-

Joint alteration	Joint water	Aperture (mm)					Joint filling					Joint spacing (m-mm)				
		>5 mm %	1 - 5 mm %	0.1 - 1 mm %	<0.1 mm %	Close	Soft filling mm 5< %	Soft filling mm 5> %	Hard filling mm 5> %	Hard filling mm 5< %	Clean %	<60 mm %	60 - 200 mm %	200 - 600 mm %	0.6 - 2 m %	>2 m %
		-	100	-	-	-	-	50	-	50	-	-	-	-	100	-
W1	Dry	100	-	-	-	-	100	-	-	-	-	-	-	100	-	-
		40	60	-	-	-	-	-	70	-	30	-	-	30	70	-
		-	100	-	-	-	70	30	-	-	-	-	-	-	-	100

Table 13. Geomechanical classification of rock masses in unit three of the Atamir Formation K (at) 2

Parameter	Value	Rating
UCS (Mpa)	40 - 60	4 - 6
RQD	25 - 50	8
Spacing (mm)	>60 - 200	8
Condition of discontinuities		10
Groundwater	Dam	7
RMR	35 - 39	
Class No.	Dam	
Cohesion of the rock mass (KPa)	190	
Friction angle of the rock mass (Deg)	15 - 25	

Table 14. The Q index of rock masses in unit three of the Atamir Formation K (at) 3

Parameter	Description	Value
RQD	37 - 43	37 - 43
J_n	Three joint sets with random to four joint sets	12
J_r	Smooth to planar	1
J_a	Softening or low friction clay mineral coatings	4
J_w	Dry to medium inflow or pressure	1 - 0.66
SRF	Medium stress, favorable stress condition	1
Q	0.51 - 0.89	
Class No.	Very poor rock	

Table 15. Geomechanical parameters of the rock masses in unit three of the Atamir formation K (at) 3

Parameter		Min	Max	Ave
GSI		30	40	35
Intact uniaxial compressive strength (MPa)		30	80	55
Mi		9	17	13
Material constants	m_b	0.457	0.754	0.366
	S	0.0001	0.0009	0.0001
	A	0.489	0.654	0.516
Rock mass parameter (MPa)	Cohesion (Mpa)	0.190	0.451	0.270
	Friction angle (Deg)	19.05	50.63	38.41
	Tensile strength	−0.021	−0.09	−0.012
	Uniaxial compressive strength	0.023	1.98	0.426
	Global strength	1.65	9.76	4.075
	Modulus of deformation	600.75	1000.5	347.07

Table 16. Geomechanical parameters of the rock masses in the three units of the Atamir Formation

Unit	RMR	Q	GSI	Factor of safety (FS) of the wedge
Unit one of Atamir Formation K (at) 1	34 - 36 Poor	0.592 - 0.66 Very poor	35 - 45 Poor	0.037
Unit two of the Atamir Formation K(at) 2	37 - 41 Poor	0.62 - 0.83 Very poor	40 - 50 Poor	0
Unit three of the Atamir Formation K (at) 3	35 - 39 poor	0.51 - 0.89 Very poor	30 - 40 Poor	0

Table 17. Geomechanical parameters of the rock masses in unit three of the Atamir Formation K (at) 3

Set number	Characteristics and parameters of rock mass	Rating rock masses				
		81 - 100 (I)	61 - 80 (II)	41 - 60 (III)	21 - 40 (IV)	<20 (V)
1	Rock classification	Very good	Good	Relatively good	Poor	Very poor
2	Average time of self supporting capacity	10 years for active opening of 15 meters	6 months for active opening of 8 meters	1 week for active opening of 5 meters	10 hours for active opening of 2.5 meters	30 minutes for active opening of 1 meter
3	Cohesion of rock mass (Mpa)	>0.4	0.3 - 0.4	0.2 - 0.3	0.1 - 0.2	<0.1
4	Angle of internal friction	>45 degrees	35 - 45 degrees	25 - 35 degrees	15 - 25 degrees	<15 degrees
5	Load bearing capacity (T/M^2)	440 - 600	280 - 440	135 - 280	45 - 135	30 - 45

Therefore, rockslide has been unavoidable in rock masses of all three units. The point of great importance is the mass movement of the rocks towards the road and channel that has blocked them and will cause floods during future rainfalls. Based on observations made, the process of landslide under discussion has not ended, and it seems that a long time has to pass (from several months to several years) for stability to be established.

Condition of the rock masses (Table 16).

Considering the RMR stability table (Table 17).

REFERENCES

1. Feenstra, R. and Wickham, J. (1975) Evolution of Folds around Broken Bow Uplift, Ouachita Mountains, Southeastern Oklahoma. AAPG Bulletin, 59, 974-985.

2. Bieniawski, Z.T. (1973) Engineering Classification of Jointed Rock Masses. Civil Engineer in South Africa, 15, 353.

3. Cai, M., Kaiser, P.K., Uno, H., Tasaka, Y. and Minami, M. (2004) Estimation of Rock Mass Strength and Deformation Modulus of Jointed Hard Rock Masses Using the GSI System. International Journal of Rock Mechanics and Mining Sciences, 41, 3-19.http://dx.doi.org/10.1016/S1365-1609(03)00025-X

Chapter 2

GROUNDWATER RISING AS ENVIRONMENTAL PROBLEM, CAUSES AND SOLUTIONS: CASE STUDY FROM ASWAN CITY, UPPER EGYPT

Sayed A. Selim[1], Ali M. Hamdan[1], Ahmed Abdel Rady[2]

[1]Geology Department, Faculty of Science, Aswan University, Aswan, Egypt

[2]Aswan Water and Sanitation Company, Aswan, Egypt

ABSTRACT

This paper examines the rise in the level of the groundwater in the Quaternary aquifer at Aswan city, Upper Egypt. Since the 1960's, the areal extent of Aswan City and the urban populations are growing at a high pace which introduces new sources of water that increase groundwater recharge. As a result of leakages or infiltrations from different sources, the natural groundwater balance is overturned into an unbalance where the input to water table is comparatively much more than the natural groundwater flow towards the Nile River. The present study shows a variation in the groundwater level, from 1971 up to 2014, where the water table rising ranges between 12.55 and 13.69 m. Also, it shows an abrupt increase in the water levels in 2010 continuing up till now. The groundwater rising phenomena that happened in 2010 can be directly refereed to the cessation of groundwater pumping from El-Shallal wells, and to the reduction of pumping from KIMA factory wells. Generally, the rate of water rising is much higher in the western side of the city and in Kima factory area, where they are characterized by low relief and dense population. The most troublesome groundwater mounds under urban areas are likely to develop in low-lying areas of relatively high permeability aquifer, which is not exploited for water supply. These damages will become more widespread if the rising groundwater table remains uncontrolled. The environmental impact of the water rising includes: forming ponds in low lying areas (Kima and El Shallal ponds), flooding building's basements, and inundating underground infrastructure. A general deterioration in groundwater quality was identified.

INTRODUCTION

The area under investigation (Aswan City) is situated along the Nile Valley in the southern part of Egypt. It is located adjacent to the east bank of the Nile River, bounded by latitudes 24°01'30" and 24°06'30"N and longitudes 32°52' and 32°56'E (Figure 1).

Over last few years, a rise in groundwater levels has been observed in several parts. Raising groundwater levels have the potential to inundate underground infrastructure, flooding basements and submerging sewer pipes and utility lines that deliver water and electricity. The magnitude of anthropic impacts upon their environment makes humans the major geologic agent on the surface of the planet [1] .

This work aims to review the causes of changes in groundwater levels in Aswan region which is under the subtropical arid region conditions. The geo-environmental impact of the rising water-level and their consequences on the built environment will be considered.

Over the last few years, a steady rise in groundwater levels has been observed in several parts of Aswan City. This reflect environmental problems existing in many areas of the city, where it creates swamps and ponds and affects the basement of many buildings as it is shown in El-Seil, KhorAwada, Phatemic graves, El-Aqad buildings, Blood Bank, Military building, El Shallal and KIMA factory area (Figure 1 and Figure 2). Rising groundwater levels are expected to be a chronic problem and will likely be a major issue for residential areas of Aswan city.

STUDY AREA

Geomorphologically, Aswan City area can be divided into three main units: low elevated lands (Aswan plain); high elevated lands; and the Nile River channel (Figure 1).

1) Low lands (Aswan plains): are underlain by the urban areas of Aswan city and can be divided into two main parts. The western plain that extends along El Sadat road up to Atlas area northward, while the eastern plain extends along El Samad road (Figure 1). These two plains were the location of the old channel of the Nile River and their width determined the original width of the old Nile gorge. The maximum width of these plains is encountered near the center of the town (about 3.5 Km), and the minimum one is along El Sadat road (less than 1 kilometer). Generally, these plains are characterized by great variation in ground elevation and show a general slope from south to north (Figure 3).

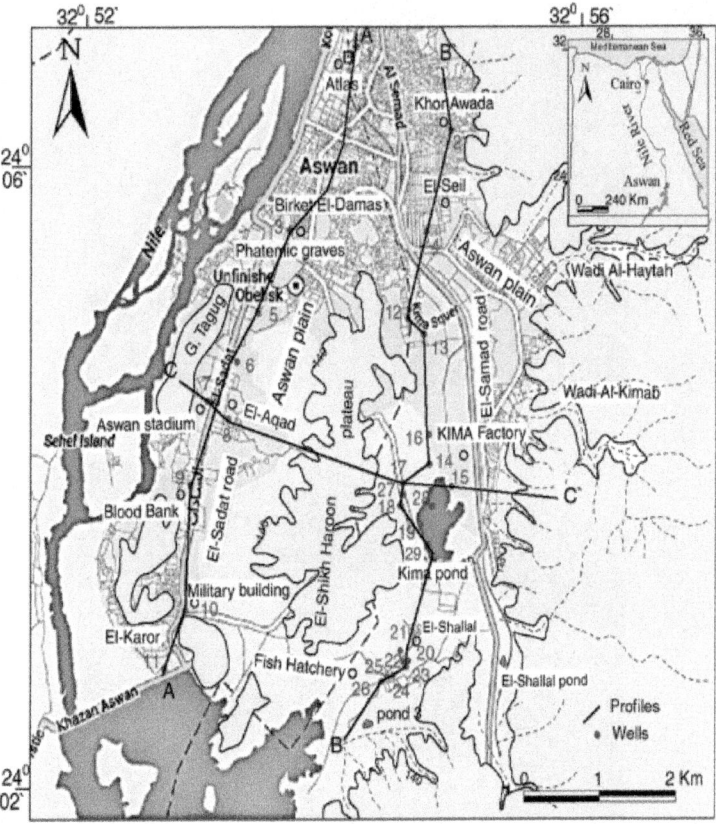

Figure 1: Location map of drilled wells and profiles in the study area.

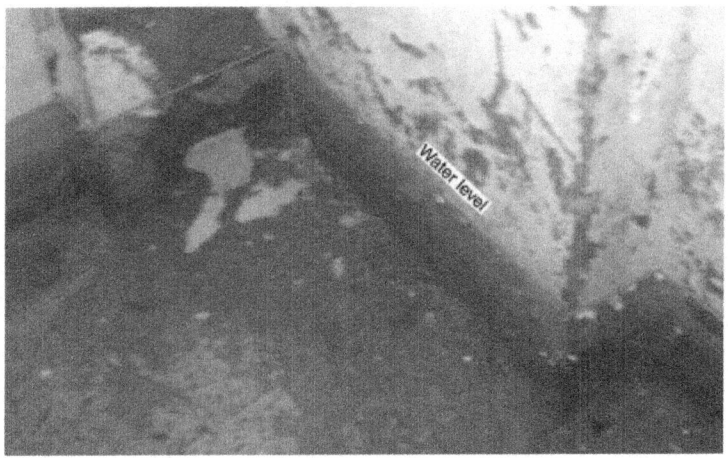

(a)

Figure 2: Rising groundwater level affecting areas: (a) KhorAwada, (b) El-Seil, (c) Phatemic graves area, and (d) Aswan stadium.

2)   The highlands: the more extensive one bounded the study area from its eastern side (elevations vary from 164 to 188 m). Few wadis incise this highland and directed from east to west, W. El-Keimab and W. Al-Haytah (Figure 1). The second main highland occupies the central parts of the study area and represented by El-Shikh Haroon plateau (elevations vary from 142 to 167 m). Other small highlands are located to west of the study area close to the Nile River as Gebel Tagug (150 m) and El Karor (135 m). The high land areas are underlain either by basement rocks and or by Nubian sandstones.

3)   The Nile River Course forms the western limit of Aswan City area (the approximately water level is +85 m). Aswan Dam (Khazan Aswan) with its water reservoir is situated to the south of Aswan city where the water level ranges between 106 and 118 m.

The general geology of Aswan area is relatively well known and has been studied by several investigators [2] - [10] .

The following lithological units were exposed in the study area (Figure 4).

•   The basement rocks represent the oldest rocks (Precambrian age) and mainly exposed in the many parts of the study area (Figure 4). They are characterize by the presence of many fissures and joints, and can play an important role in accumulation of water in many parts of the study area.

•   The Nubian Sandstone rocks of Cambrian to cretaceous age [11] are exposed in the central and eastern parts of the study area.

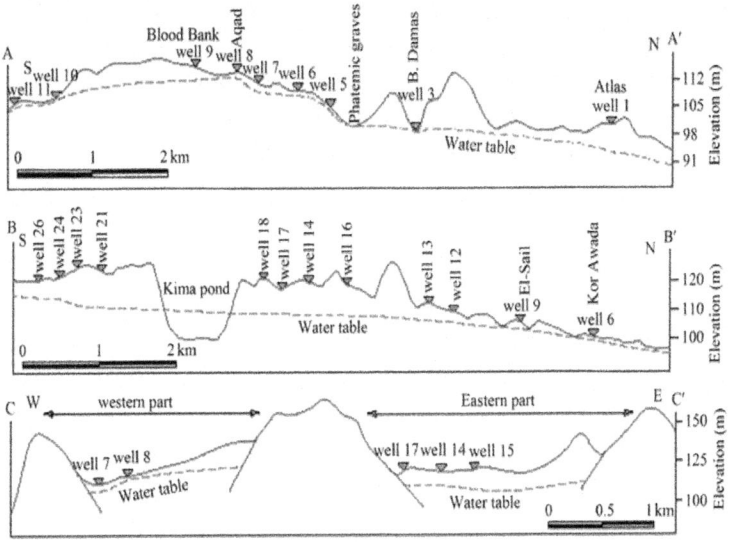

Figure 3: Hydrogeological profiles along the study area (2014).

It exposed at high lands in the central part (El Shikh Haroon plateau) and the eastern part and unconformably lies over the basement complex (Figure 4). It is made of ferruginous sandstone, sandstone and clays and composed into three Formations from base to top, Abu Agag Formation; Timsah Fomation; and Um Barmile Formation and has a total thickness ranges between 20 and 85 meters [12] .

- The Quaternary sediments are represented by sands, gravels and clays of the Pleistocene time, and by mud and Aeolian sediments of the recent. They are underlain by a thick bed of Pliocene clays [13] . These sediments are well exposed in the area through El-Samad and El Sadat roads with a thickness reached up to 248 m. The Quaternary gravels are composed of coarse well-rounded pebbles, ranging from 15 to 20 centimeters in diameter. The recent deposits comprise a small portion of the surface of the study area and are represented by alluvial and Aeolian deposits. The alluvial deposits, sand and mud form the cultivated land in the northern part of the study area. The Aeolian deposits are represented by blown-sand.

Hydrogeologicaly, few has been published on the hydrogeology of Aswan city area due to the limited amount that groundwater used for public water supply. The Quaternary sediments represent the main aquifer in the studied area and outcropped in two geomorphologic low lands (Aswan plains) which will be outlined in the following.

At the western plain (along El Sadat road): The quaternary aquifer is consists of unconsolidated sediments of sands, gravels, and clays enriched of smectite minerals group of the Pleistocene time. These sediments are unconformably overlies the basement rocks and fill the floor of the low lands with a thickness vary from few meters to less than 20 m. basement rocks outcrops in many parts along the western part of the study area as in El Karor and G. Tagug.

The eastern border of the aquifer is El-Shikh Haroon plateau (basement rocks caped by The Nubian Sandstone rocks), but westward is bounded by G. Tagog and El Karorplateaux. The width of the aquifer is relatively small and varies from 1.5 km (at Blood Bank) to 1.67 km (at El-Aqad). It reached 4.52 km at the central part of the city and 1.13 km at the northern extreme (Atlas).

At the eastern plain (along El-Samad road): The quaternary aquifer in the eastern part of the study area studied by [14] [15] . They concluded that the Quaternary aquifer in the eastern part is mainly composed of unconsolidated material of sands, gravels, and clays intercalation. It is not covered by impermeable layers in the major part of the area; therefore, it is under unconfined condition. They stated that the transmissivity (T) values of the Quaternary aquifer range between 1996.4 and 3029 m^2/day which means that the aquifer is of high potential class (more than 500 m²/day) according to the classification of [16] .

The thickness of the Quaternary aquifer vary from 100 m in the southern part to 137 m in the central part and it underlined by the Pliocene clay, as detected from the lithological log of deep well No. 27 (Figure 5). The east- ern border of the quaternary water-bearing sediments represented by the basement rocks which caped with Nubian sandstone, while westward is bounded by high land of El-Shikh Haroon plateau.

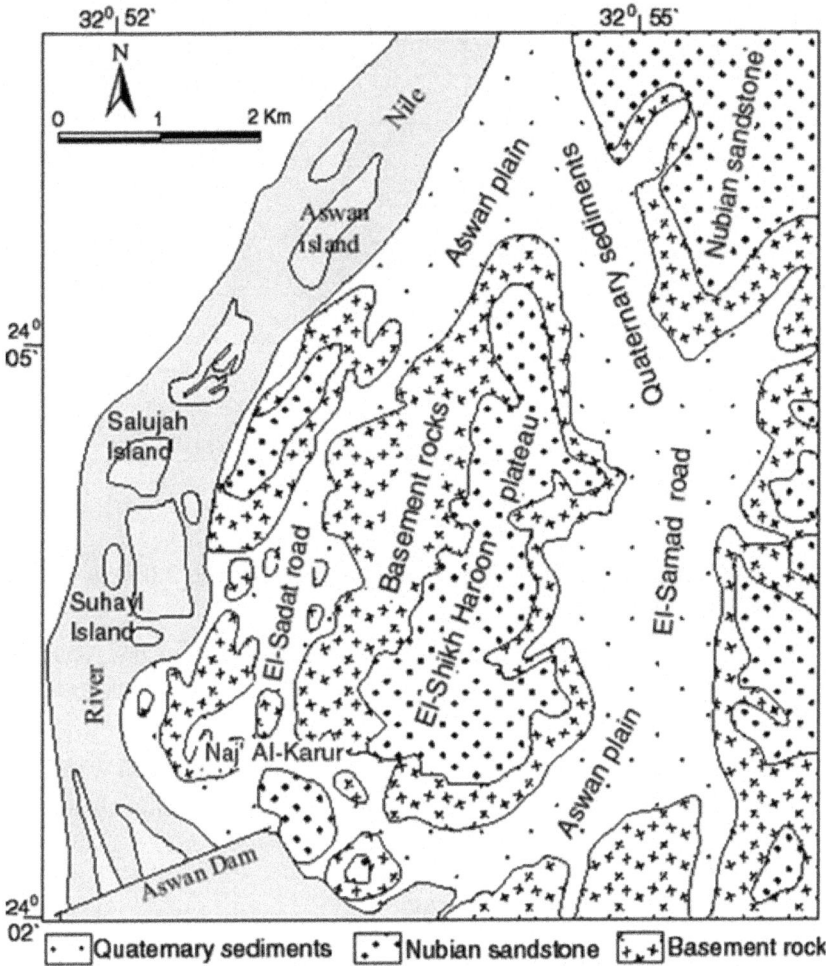

Figure 4: Geological map of the study area.

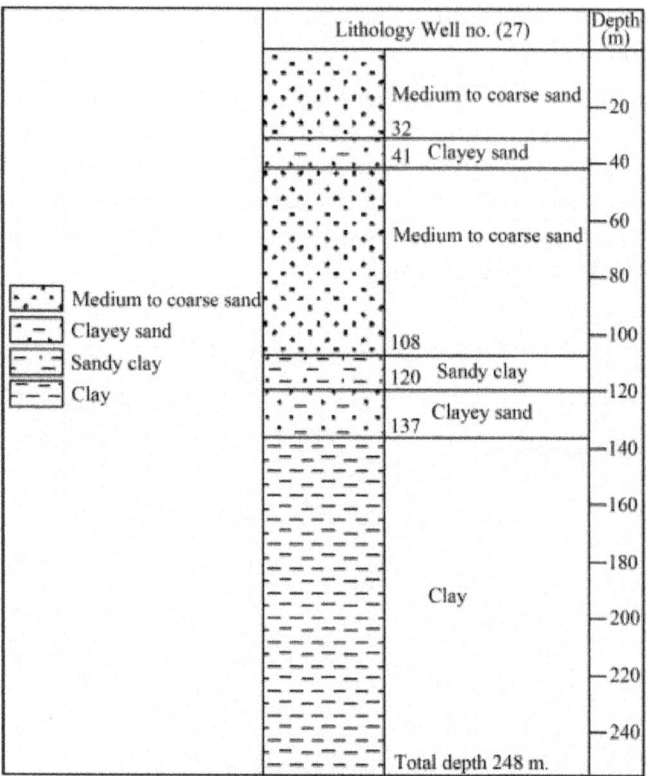

Figure 5: Lithological log of well No. 27.

The width of the aquifer is small and ranges between 2.05 km northward and 1.37 km southward.

The natural recharge to the Quaternary aquifer is from the Aswan Dam Lake, which has water level ranges between 108 and 118 m asl, towards the urban areas of the city.

The aquifer discharged through the natural groundwater outflow to the Nile River, in the northern part of the city in addition to groundwater abstraction by pumping at Kima and El Shallal areas.

MATERIALS AND METHODS

During this study, the data recorded from 1971 to 2014, from six wells at Kima area are evaluated to determine the general trend in groundwater levels and the change in the water-table over a 43-year period (Table 1 and Figure 6). These data also helped to identify general and local groundwater flow directions, where the groundwater level contour map is prepared (Figure 7).

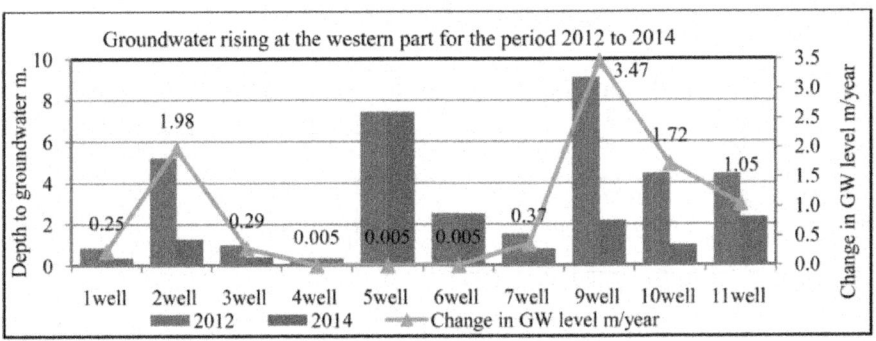

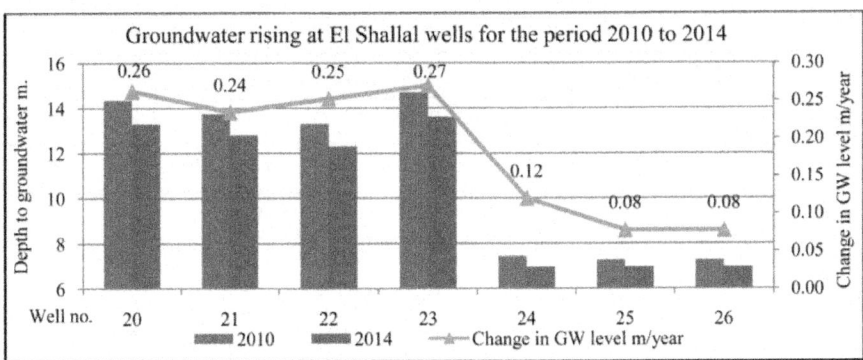

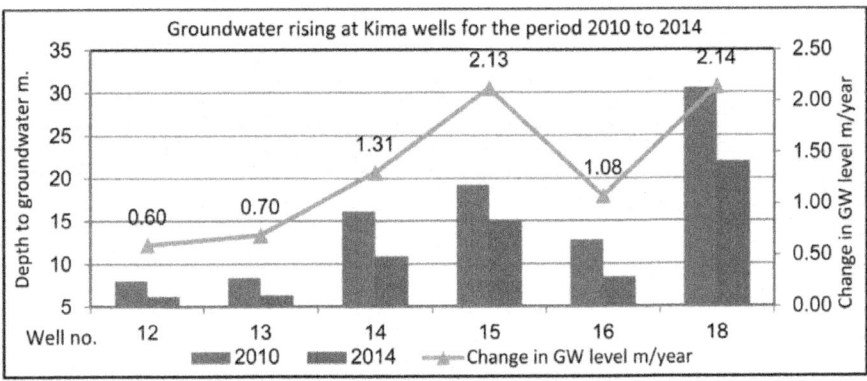

Figure 6: Groundwater rising and changes in GW level per year in the study area.

Table 1: The hydrogeological data of drilled wells of the study area

Date	Water depth	Water Level	Date	Water depth	Water Level	Date	Water depth	Water Level
Well No. 1			Well No. 26			Well No. 13 (continue)		
1/6/2012	0.85	101.86	18/7/2010	7.26	113.74	20/10/2012	6.49	114.51
22/2/2014	0.36	102.35	15/1/2014	6.95	114.05	3/11/2012	6.42	114.58
Well No. 2			Well No. 12			1/12/2012	6.37	114.63
1/6/2012	5.21	104.74	17/10/1971	19.0	91.0	21/1/2013	6.42	114.58
22/2/2014	1.25	108.70	26/6/1975	18.75	91.25	16/2/2013	6.4	114.6
Well No. 3			16/1/1978	15.5	94.5	16/3/2013	6.36	114.64
1/6/2012	0.98	99.01	14/8/1979	15.9	94.1	13/4/2013	6.33	114.67
22/2/2014	0.4	99.59	5/1/1980	15.4	94.6	25/5/2013	6.31	114.69
Well No. 4			18/11/1981	16.1	93.9	Well No. 14		
1/6/2012	0.35	91.82	2/1/1982	14.6	95.4	24/11/1981	26.0	93.0
22/2/2014	0.44	91.73	28/7/1984	16.3	93.7	3/1/1982	25.0	94.0
Well No. 5			2/4/1985	15.6	94.4	25/10/1983	26.0	93.0
1/6/2012	7.42	84.86	31/5/1986	16.3	93.7	2/7/1984	27.0	92.0
22/2/2014	8.23	84.05	1/10/1987	16.05	93.95	1/4/1985	24.57	94.43
Well No. 6			25/1/1988	15.47	94.53	29/9/1987	22.6	96.4
1/6/2012	2.53	98.44	18/1/1989	14.9	95.1	24/1/1988	22.0	97.0
22/2/2014	2.52	98.45	14/3/1990	14.7	95.3	8/2/1989	22.8	96.2
Well No. 7			21/9/1991	13.33	96.67	9/12/1989	23.2	95.8
1/6/2012	1.53	97.05	17/12/1992	11.94	98.06	12/2/1991	22.23	96.77
22/2/2014	0.8	97.78	16/5/1993	11.77	98.23	5/2/1992	21.35	97.65
Well No. 8			15/2/1999	11.8	98.2	18/12/1993	20.67	98.33
1/6/2012	8.76	95.56	6/2/2001	12.0	98.0	5/8/1996	21.65	97.35
22/2/2014	closed	104.32	8/3/2004	11.8	98.2	11/7/2001	22.0	97.0
Well No. 9			15/2/2010	8.0	102.0	19/5/2010	16.1	102.9
1/6/2012	9.1	90.03	19/5/2010	7.5	102.5	28/6/2010	15.98	103.02
22/2/2014	2.17	96.96	28/6/2010	7.4	102.6	5/7/2010	15.87	103.13
Well No. 10			8/1/2011	6.28	103.72	28/4/2011	13.68	105.32

1/6/2012	4.44	94.97	21/6/2011	6.45	103.55	15/9/2011	13.0	106.0
22/2/2014	1.0	98.41	Well No. 13			3/1/2012	12.4	106.6
Well No. 11			6/6/1971	20.0	101.0	7/2/2012	12.9	106.1
1/6/2012	4.43	109.91	9/2/1985	16.27	104.73	25/8/2012	14.55	104.45
22/2/2014	closed	114.34	17/11/1986	17.02	103.98	20/10/2012	14.37	104.63
Well No. 20			24/9/1992	12.48	108.52	16/2/2013	13.93	105.07

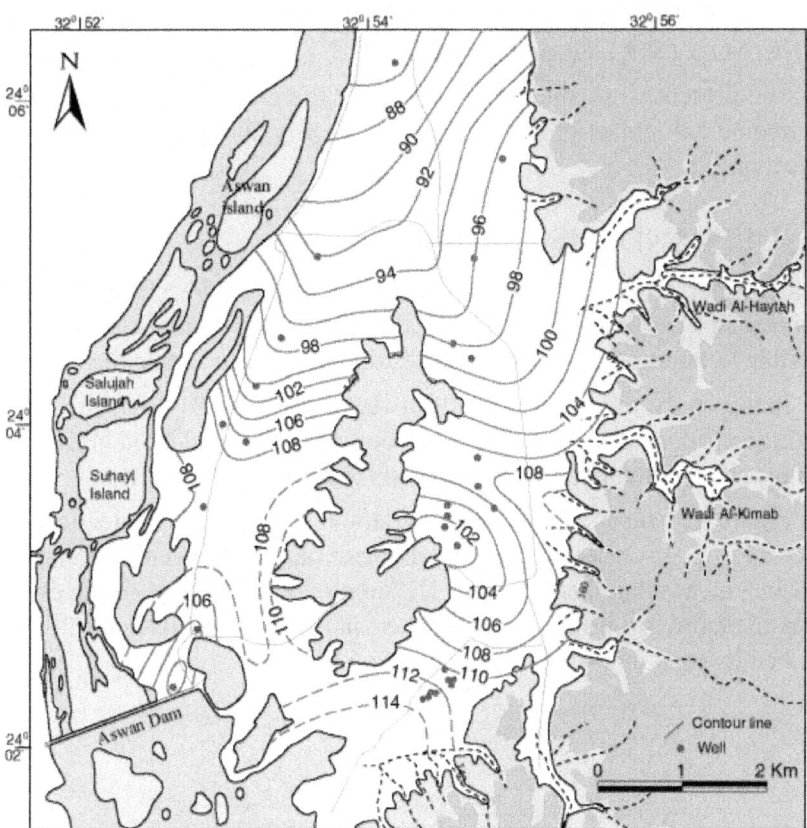

Figure 7: Groundwater level map in the study area (2014).

In order to assess the rate of groundwater rise, the present authors measured the hydrogeological data (depth to water, groundwater level, ground elevation, and total depth) from the drilled wells through many trips from 2010 up till now (Table 1). The collected data covered all affected areas through monitoring the cases of groundwater rising and their environmental impact and the illustrations were prepared (Table 1 and Figure 8, Figure 2, and Figures

9-14).

To achieve the causes of this problem in Aswan city, hydrogeological data were collected from 11 wells drilled along the study area for the period from 1971 to 2009 (Figure 15).

A total of 13 water samples were collected from some of affected areas and bore wells from various localities of the city and chemical and microbiological analyses carried out to assess the groundwater quality. They analyzed for different physio-chemical parameters (TDS, pH, E.C, TH), major cations (Ca^{2+}, Mg^{2+}, Na^+, K^+), major anions $(HCO_3^-, SO_4^{2-}, Cl^-)$, and some trace ions as (PO_4), (NO_2), (NO_3), Fe, and Mn (Table 2).

The present work studied the causes of the groundwater level rising, the environmental impact of the problem on the city and gave recommendations for solving it.

RESULTS AND DISCUSSION

Groundwater level rise results provide a basis for the characterization of groundwater variation within the Quaternary aquifer in the studied area. It is possible to interpret different processes that may occur within the aquifer.

In this study, the data recorded between 1971 and 2014, from five wells at Kima area, are evaluated in order to determine the general trend in water levels and the change in the water-table over 43-year period (Table 1).

Figure 15 shows a noticeable variation in the groundwater level, from 1971 up to 2014, where the water table rise ranges between 12.55 and 13.69 m which means that there is about 1.37 m groundwater table rise in each year. Also, the figure shows an abrupt increase in the water levels in the 2000's and reached its peak in 2010 and continuing up till now.

(a)

(b)

Figure 8: Rising groundwater level creating ponds: (a) Kima pond, (b) El Shallal pond.

Figure 9: Increase population and Urbanization activity at the study area.

Figure 10: Rash of sewage network system on streets.

From the groundwater level contour map for the year 2014 (Figure 7) the following points can be identified.

- The general groundwater flow direction is generally from the south towards the north, i.e. toward the Nile River, Other direction of water flow comes from the highlands, which is characterized by high urbanization.

- The highest water level is recorded at El Shallal area, close to the area of recharge (Aswan Dam Lake), and the lowest water level is noticed at the northern part of the city, close to the area of discharge, the Nile river.

- The contour lines are more or less regular in the northern part of the city, while they are condensed in El Shallal, Kima and El Aqad areas where these areas are characterized by dense population and industrial activities.

(a)

(b)

Figure 11: Satellite image for KIMA factory pond before flooding (a) 2009 and after five years of flooding (b) 2014.

Figure 12: Salt crystals cause the building materials to split, flake and crack (Khor Awada and El Sail areas).

- Two water depressions are noticed in Kima and El Karor areas, these two areas are characterized by low ground elevations.

The authors measured the depth to groundwater in 26 piezometers, all over Aswan city, for the period from 2010 up to 2014 to assess the rate of the water rising in the city (Table 1 and Figure 6).

(a)

(b)

Figure 13: Examples from different locations where algae growth in the rising water at the affected areas. (a) Khor Awada; (b) El Sail.

(a)

(b)

Figure 14: Groundwater rising and their environmental impact: (a) Displacement occurs between two buildings, (b) Example of flooding buried basements of many buildings. (a) El Aqad; (b) Aswan stadium.

The table and figure show that the rise in groundwater level ranges between 8 cm/year (wells 25 & 26) and 27 cm/year (well 23) at El Shallal area, and it ranges between 0.60 cm/year (well 12) and 2.14 m/year (well 18) at Kima area, while it reached 3.46 m/year (well 4) in the western plain and 0.01 cm/year (well 1) at the northern side of the city (Figure 6). We can concluded that the rate of the water rising is much higher in the western side of the city, along El Sadat road, and in Kima factory areas where they are characterized by low relief and by dense population. The minimum rising of the groundwater is noticed in the northern part of Aswan city because this area is of low population and represents the discharge area for the Quaternary aquifer; to the Nile River. During the last five years of monitoring, some wells at KIMA factory area are submerged under the rising water (wells No. 19, 28 and 29).

In 2009, 40 productive wells at El Shallal area were turned off, as a result of deteriorated water quality, they were pumped an amount of 12.44×10^6 m³/year. Moreover, groundwater withdrawals from KIMA factory wells were reduced from 13.23×10^6 m³/year to 9.12×10^6 m³/year. The former discussion proves that the phenomena of groundwater rising that happened in 2009 at Aswan city, can be directly refereed to the cessation of groundwater pumping from El-Shallal wells, and to reduction of pumping from KIMA factory wells.

Causes of the Water Level Rising Problem

The problem of increasing the groundwater level and flooding over the ground surface is observed in many lowlands along the study area as El-Seil, Khor Awada, Birket El-Damas, Phatemic graves, El-Sadat road, El-Aqad, Aswan stadium, Blood Bank, Military building, Kima, El Shallal (Figure 1 and Figure 2). The main causes of the water rising at the Aswan city can be related to the following.

Table 2: Chemical analysis of the surface and groundwater samples in the study area

Sample No.	Physio-chemical parameters				Major Cations (ppm)				Major Anions (ppm)			Trace ions (ppm)					e%
	pH	EC μmho/cm	TDS (ppm)	TH (ppm)	Ca	Mg	Na	K	HCO_3	SO_4	Cl	(PO_4)	NO_2	NO_3	Fe	Mn	
well 1	7.65	540	454	231.9	50	26	16	7.4	275	22.8	15.2	0.11	0.1	3.52	0.016	0.005	0.93
well 2	7.86	5250	3383	993.8	277	73.4	700	14.5	908	550	860	0.13	0.15	3.74	0.011	0.006	0.07
well 3	7.9	8800	5632	1482	424	103	1320	50	360	1650	1700	0.15	0.14	0.19	0.012	0.004	0.05
well 4	8.4	6510	4166	1198	320	97	925	20	340	1382	1068	0.19	1.12	86.4	0.151	0.053	0.15
well 5	7.45	4760	3046	900.2	250	67	650	14.3	320	1042	690	0.17	0.13	0.16	0.017	0.006	0.21
well 6	7.84	4360	2812	833.3	248	52	620	14.8	321	900	657	0.177	0.16	14.3	0.537	0.036	1.66
well 7	8.36	1170	748	270.1	62	28	120	7.4	200	128	160	0.21	0.13	0.19	0.173	0.044	1.58
well 9	7.6	2530	1619	442.3	98	48	340	19	395	332	375	0.19	0.17	0.18	0.113	1.038	0.29
well 10	7.63	3940	2568	522.6	145	39	640	26	380	685	653	0.18	0.17	0.14	0.02	0.006	0.03
well 11	7.39	1250	845	338.8	78	35	135	7.4	265	155	170	0.2	0.18	8.7	0.02	0.004	1.80
Pha-temic graves	7.85	9150	5856	3146	530	443	603	42.3	1827	1181	1255	3.49	0.08	2.12	0.085	0.09	0.13
Khor Awada	8.09	4420	2828	763	106	121	636.4	37.3	792	538	597.5	0.81	0.37	15.8	0.18	0.21	3.33
Aqad	8.03	3810	2418	659	114	91	530.8	35.8	628	456	562	0.72	0.48	14.1	0.11	0.25	2.09

Variation of Direct Groundwater Recharge and Discharge

The principal cause is the great variation of the difference between the recharge and discharge to and from the main aquifer under Aswan city area, where a big and continues recharge amount of water regardless to a small discharge amount of water by time lead to increasing the water level and appear in low elevated lands in the study area causing a concern problem. The groundwater recharge from potable and different supplies may exceed both the natural rate of recharge from Aswan Dam Lake and the natural rate of groundwater discharge. So that, less discharge of groundwater indirectly leads to increase the water level.

The Hydrogeological Parameters of the Quaternary Aquifer

The hydrogeological parameters of the Quaternary aquifer, in Aswan area, play significant role in the calculations of groundwater rise. The pumping tests analyses reveal that the Quaternary aquifer at Aswan area has high hydraulic conductivity (ranges between 0.0001 and 0.0004 m/s), and high transmissivity values (0.02 to 0.04 m²/s), [14] . Increasing hydraulic conductivity and transmissivity lead to increasing the rate of groundwater flow which act as an important role in the problem of water rising.

Cessation of Groundwater Pumping

In 2009, 40 productive wells at El Shallal area were turned off, as a result of deteriorated water quality, they were pumped an amount of 12.44×10^6 m³/year. Moreover, groundwater withdrawals from KIMA factory wells were reduced from 13.23×10^6 m³/year to 9.12×10^6 m³/year. This proves that the phenomena of groundwater rising that happened in 2009 at Aswan city, can be directly refereed to the cessation of groundwater pumping from El-Shallal wells, and to reduction of pumping from KIMA factory wells.

Groundwater Recharge from the Urban Activities of High-Land

Urbanization has a great effect on the groundwater regime especially to subsurface components of infiltration and percolation leading to groundwater recharge.

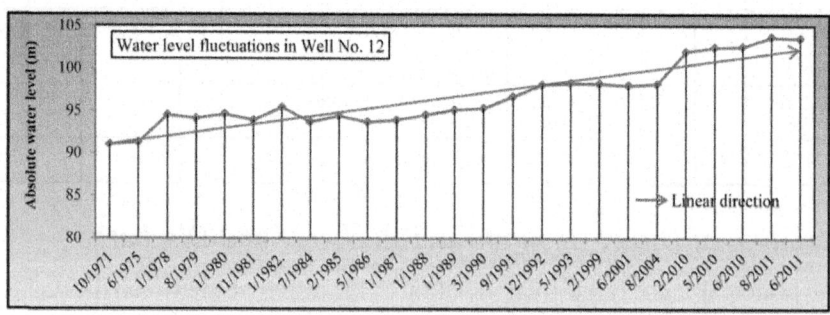

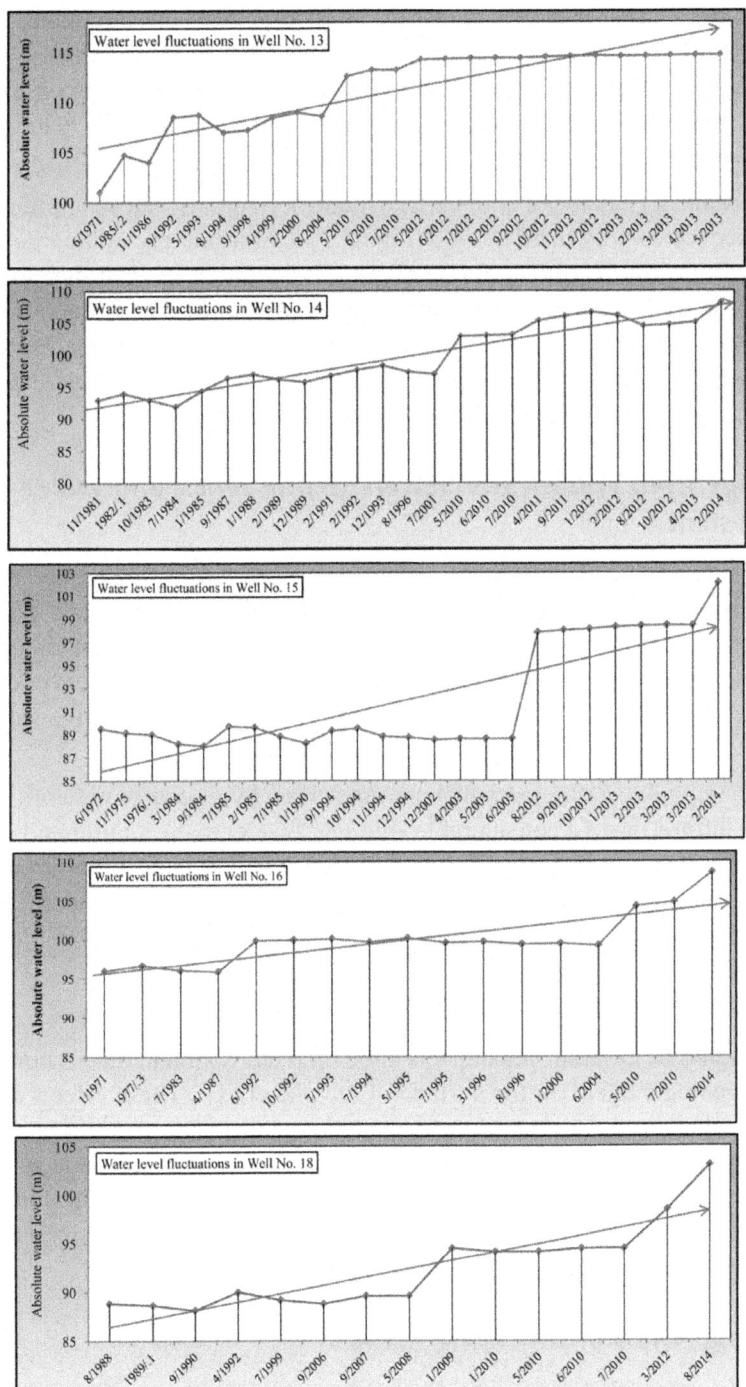

Figure 15: Water level fluctuations in different wells from 1971 till now.

In any urbanization, sanitation and drainage arrangements are of fundamental importance. Quantifying groundwater recharge in urban areas is especially challenging because the urban environment is quite complex as a large variety of land uses coexist and overlap and because of the heterogeneity of the shallow underground.

Aswan city expanded both vertically and horizontally due to unprecedented rapid economic growth after 1970 (Figure 9). Unfortunately, neither geological nor hydrological features are taken into consideration in such developments. Water supply from alternative sources and their distribution through the mains cause to water leakages within urban areas, which enhance groundwater table rise. Moreover, the aquifer beneath the urban area is unexploited which leads to continuous groundwater level rising under these areas.

Leakage from Wastewater Sewers, Septic Tanks, and Water Supply System

When sewer lines are located below the water table, they may infiltrate groundwater, and when located above the water table they may leak. Therefore, the seepage from sewage network system itself conceder another problem. Recently in Aswan city area a phenomena of rash of sewage network system on streets (above ground surface) for many times leads to raise the water level and become as other recharge source (Figure 10).

Septic tanks, in the parts that are not served by a sewage network system, lead to infiltration of a considerable amount of water to the aquifer and increase the water level. In many parts of the study area, seepage water from water supply networks, even if small amount, give another reason of water rising and flow water to accumulate and appear in low lands.

Environmental Impacts of Water Level Rising

The magnitude of anthropic impacts upon their environment makes humans the major geologic agent on the surface of the planet [1] . These effects are most severe where population concentrates, today, half of the world's populations live in urban areas. Water quality is a prime issue in urban settings as shallow aquifers and surface waters in cities are subject to pollution by a multitude of point and non-point sources. At Aswan city, the environmental impact of the water level rising includes the following points.

Forming Ponds in Low Lying Areas

In many areas of Aswan city, the continuous increase in the groundwater rising leads to creation of new ponds e.g. Kma, El Shallal, and pond 3 (Figure

1 and Figure 8). These ponds have bad environmental impact to the area, where theymay threaten the community public health. At KIMA factory area, the continuous rising of groundwater level leads to increasing its area and covering new lands (Figure 11).

Impact on the Groundwater Quality

A total of 13 water samples were collected from some of affected areas and bore wells from various localities of the city and chemical analyses carried out to assess the groundwater quality.

They analyzed for different physio-chemical parameters (TDS, pH, E.C, TH), major cations (Ca^{2+}, Mg^{2+}, Na^+, K^+), major anions $(HCO_3^-, SO_4^{2-}, Cl^-)$, and some trace ions as (PO_4), (NO_2), (NO_3), Fe, and Mn (Table 2).

In the study area, the average value of TDS is 2798.1 mg/l (Table 2). The relatively high TDS of the analyzed water samples have a bad environmental impact on the infrastructures of the area. The capillary action draws the salty groundwater to the surface and into the porous walls and foundations of the buildings. The dry desert heat causes the evaporation of the water from the walls material leaving behind salt crystals that cause split, flake and crack of these walls (Figure 12).

The salt crystals combinations are calculated for both the surface and groundwater samples along the study area. They revealed the presence of different groups of salt assemblages (Table 2). The hypothetical salt combination revealed the presence of different salts arranged in terms of their predominant as $NaCl$, $Ca(HCO_3)_2$, Na_2SO_4, $MgSO_4$, $CaSO_4$, $Mg(HCO_3)_2$, and KCL; where the average of equivalent percentage is 42.18%, 19.16%, 13.68%, 10.42%, 9.01%, 4.34%, and 1.37% respectively. In many spots of the study area, high nitrate, sulphate, chloride and sodium concentrations indicate anthropogenic aquifer pollution and it is most severe where population concentrates.

Growth of Micro-Organism and Algae

Some of the collected water samples were subjected to microbiological measurements for the total bacteria and total Coliform in the study area (Table 3). The total Bacteria vary from 20 to 1000 no of colony/100 ml. The total Coliform ranges between 0 and 470 no. of colony/100 ml. In Phatemic graves and KhorAwada, the total bacteria and total Coliform are too numerous to count. The results obtained from the microbiological analysis reflect a presence of bacterial activity in the accumulated water due to rising groundwater level in the study area.

Once a building has been exposed to a large volume of water, may species of micro-organism and algae can growth (Figure 13). They are causing bad environmental impact on building and human health. In addition of directly environmental effect on building and human health of pollutants water of micro-organism, the probability of pollutants water to seeps to the drinking fresh water pipes network increase.

Flooding Building's Basements and Inundate Underground Infrastructure

In many locations of the study area, water level rising leads to inundation of the basement of many buildings (Figure 2 and Figure 14(b)). In any urban area, presence of the groundwater for long time has bad environmental impact on building, where it causes direct damage to building. Figure 14(a) shows clear displacement occurs between the two building as result of water level rising and flooding their basements for long time.

Table 3: The analytical of total bacteria and total Coliform in some water samples.

Sample No.	Total Coliform (colony/100 ml)	Total Bacteria (colony/100 ml)	Sample No.	Total Coliform (colony/100 ml)	Total Bacteria (colony/100 ml)
well 2	0	20	well 10	470	1000
well 4	0	20	well 11	20	40
well 5	10	50	Aqad	273	400
well 6	250	450	Phatemic graves	Too Numerous to count	Too Numerous to count
well 7	0	500	KhorAwada	Too Numerous to count	Too Numerous to count

Underground infrastructures that lie beneath many parts in Aswan City, as Communication networks, wastewater sewers system, pipes of drinking water distribution networks, high voltage electrical cables, and others, are flooded by rising groundwater level.

CONCLUSIONS

Aswan city expanded both vertically and horizontally due to unprecedented rapid economic growth after the year of 1970. Unfortunately, neither geological nor hydrological features are taken into consideration in such developments.

As a result of leakages or infiltrations from different sources, the natural groundwater balance is overturned into an unbalance where the input to water table is comparatively much more than the natural groundwater flow.

In this study, the data recorded between 1971 and 2014, are examined in order to determine the general trend in water levels and the change in the water-table over 43-year period. It shows a noticeable variation in the groundwater level during this period, and the water table rising ranges between 12.55 and 13.69 m, which means that there is about 1.37 m groundwater table rise in each year. Recently, an abrupt increase in the water levels is noticed and reached its peak in 2010 continuing up till now. This study proves that the phenomena of groundwater rising can be directly refereed to the cessation of groundwater pumping from El-Shallal wells, and to the reduction of pumping from KIMA factory wells. During these last five years, all the low-lying areas of Aswan city are suffering from groundwater rising. The continuous increase of groundwater level can lead to appearance of new affected areas in other parts at Aswan city.

The environmental impact of the water rising includes: forming ponds in low lying areas (Kima and El Shallal ponds), flooding building's basements, and inundating underground infrastructure. A general deterioration in groundwater quality was identified. Much higher sulphate and alkali and alkaline earth metal concentrations were found in many spots of the study area and it is most severe where population concentrates. Furthermore, the bacteriological investigations show that the total Bacteria vary from 20 to 1000 no. of colony/100 ml and the total Coliform range between 0 and 470 no. of colony/100 ml indicating that local sources are strongly influencing the observed bacteriological data. In some locations (Phatemic graves and KhorAwada), the total bacteria and total Coliform are too numerous to count.

RECOMMENDATION

The necessary recommendations to decrease and/or prevent water rising problem at Aswan City include re-exploitation of groundwater from El Shallal and Kimawells. Leakages from the networks must be controlled and necessary maintenances are obtained. In the future continuous monitoring of the water level in all the available drilled wells should be taken into consideration where, the continuous increase of groundwater level can lead to appearance of new affected areas in other parts at Aswan city. Moreover, all urbanization activities in the high elevated lands should be under control to decrease water seepage and/or flow to the low elevated lands.

REFERENCES

1. Heiken, G., Fakundiny, R. and Sutter, J. (2003) Earth Science in the City. American Geophysical Union Special Publication Series, 56, 440 p. http://dx.doi.org/10.1029/056SP

2. Attia, M.L. (1954) Deposits in the Nile Valley and the Delta. Geological Survey, Egypt, Cairo, 356 p.

3. Said, R. (1962) The Geology of Egypt. Elsevier Publishing Company, Amesterdam, 337 p.

4. Said, R. (1981) The Geological Evolution of the River Nile. Springer-Verlag, New York, 151 p.

5. Butzer, K.W. and Hanson, C.L. (1968) Desert and River in Nubia. University of Wisconsin Press, Madison, 562.

6. El Ramly, I.M. (1973) Geomorphology, Hydrogeology, Planning for Groundwater Resources and Land Reclamation in Lake Nasser Region and Its Environs. Report from Desert Research Institute to Lake Nasser Development Center Aswan, Cairo.

7. El Shazly, E.M., Adel Hady, M.A., El Ghaay, M.A. and El Kassas, I.A. (1974) Geological Interpretation of ERTS-1 Satallite Images for West Aswan Area, Egypt. Proceeding of the Ninth International Symposium on Remote Sensing of Environment, Michigan, 15-19 April 1974, 119-131.

8. Van Houten, F.B. and Bhattacharyya, D.P. (1979) Late Cretaceous Nubia Formation at Aswan South Eastern Desert, Egypt. Annals Geological Survey, IX, 408-431.

9. Issawi, B. (1981) Geology of the Southwestern Desert of Egypt. Geological Survey, Egypt, Cairo, XI, 57-66.

10. Barber, W. and Carr, D. P. (1981) Water Management Capabilities of the Alluvial Aquifer System of the Nile Valley, Upper Egypt. Technical Report No. 11, Water Master Plane, Ministry of Irrigation, Cairo, Egypt.

11. Issawi, B. and Jux, U. (1982) Contribution to the Stratigraphy of the Paleozoic Rocks in Egypt. Geological Survey, Egypt, Cairo, Vol. 64, 1-28.

12. Klitzsch, E.H. and Wycisk, P. (1987) Geology of the Sedimentary Basins of Northern Sudan and Bordering Areas. Berliner Geowissenschaftliche Abhandlungen, Reihe A, 75, 97-136.

13. RIGW (1988) Hydrogeological Map of Egypt. First Edition, Scale 1:2000,000, Research Institute for Groundwater, Ministry of Public Works and Water Resources, Cairo, Egypt.

14. Selim S. (1995) Geological and Hydrogeological Studies of the Quaternary Aquifer in Aswan Town Area, Egypt. Egyptian Journal of Geology, 39, 631-645.

15. Hamdan, A. and Abdel Rady A. (2013) Vulnerability of the Groundwater in the Quaternary Aquifer at El Shalal-Kema Area, Aswan, Egypt. Arab Journal of Geoscience, 6, 337-358.http://dx.doi.org/10.1007/s12517-011-0363-y

16. Gheorhge, A. (1979) Processing and Synthesis of Hydrogeological Data. Abacus Press, Preston, 390 P.

Chapter 3

FEASIBILITY OF GROUNDWATER BANKING UNDER VARIOUS HYDROLOGIC CONDITIONS IN CALIFORNIA, USA

Saad Merayyan[1], Samsor Safi[2]

[1]California State University, Sacramento, CA, USA
[2]Sacramento Area Sewer District, Sacramento, CA, USA

ABSTRACT

This study evaluates the feasibility of groundwater banking in the Central Basin. The Central Basin is located in Sacramento County in northern California, USA. The study basin is bounded by three rivers (the Sacramento, the American, the Consumes and Mokelumne rivers), and by the Sierra-Nevada mountain range. This study focuses on the potential for groundwater recharge in the Central Basin for three water years (critical, wet, above normal). For that purpose, a 3-D Groundwater Modeling System (GMS) with MODFLOW was created. Three recharge wells were added to the calibrated groundwater model to recharge the water table with 10,000 Acre-Feet (AF) of water to the Central Basin. The banking of 10,000 AF during the critical and wet years was effective in raising the water table elevation in the cone of depression area without causing any negative impact elsewhere in the basin. According to the findings of the Central Basin model, banking up to 10,000 AF of groundwater during any year type is feasible. More than 10,000 AF of groundwater banking might cause more negative impacts than positive benefits.

INTRODUCTION

This project investigates the feasibility of groundwater banking in the Central Sacramento County Basin (Central Basin). The area of the Central Basin overlies the California State Department of Water Resources' (DWR) Groundwater Basin Number 5-21.65, the South American Sub-basin. There are some differences between the Central basin size and boundaries to that of the DWR Groundwater Basin Number 5-21.65. The reason for the differences in the boundaries is due to the fact that the Central Basin was developed based on the Sacramento County IGSM Grid. [1] . For the purpose of this study, the area of DWR's South Sub-Basin American sub-basin was chosen due to its natural boundaries of three perennial rivers. The Central Basin is defined as the area bounded on the west by the Sacramento River, on the north by the American River, on the south by the Consumes and Mokelumne rivers, and on the east by the Sierra-Nevada mountain range. Figure 1 presents the general location of the area Central Basin. The surface area of the Central Basin is 388 square miles (248,000 acre). The average annual precipitation in the basin ranges from about 14 inches along the western boundary to greater than 20 inches along the eastern boundary [1] . The eastern basin boundary is defined by the uprising foothills of the Sierra Nevada mountain range. This represents the approximate edge of the alluvial basin, where little groundwater flows into or out of the groundwater basin from the Sierra Nevada foothills [2] . The western portion of the Central Basin consists of nearly flat floodplain deposits from the Sacramento, American, and Consumes rivers. The DWR publication Bulletin 118-3, 1974, indicated that the groundwater movement in the Central Basin occurs in a shallow aquifer underlying by a deeper aquifer [3] .

The shallow aquifer extends approximately 200 to 300 feet below the ground surface. While the base of deep aquifer averages approximately 1400 feet below the ground surface. The deep aquifer is separated from the shallow aquifer by a discontinuous clay layer, which serves as a semi-confining layer for the deep aquifer. The Central Basin aquifer generally receives its recharge from the rivers around the basin and precipitation infiltration. There might be some interaction with other basins but at very deep elevations [4] . Groundwater elevation in the Central Basin has been declining since the 1950s, with some recovery during the mid-1990s [1] . The recovery in 1990s was attributed to the increase use of surface water. Since then, the southern and central portions of the basin have an increased dependence on groundwater. Due to this reason, there has been a continuous trend of declining groundwater levels and a formation of a cone of depression. The groundwater levels generally vary between 10 and 90 feet below mean sea level in this south-central portion of the Central Basin, whereas conditions in the west and north have been stable [1] .

The Central Basin's potential average groundwater extraction rate is 273,000 AF/year with the extractions rates currently close to upper extraction limits. With the continuous increasing demand on water resources in the region, the Central Basin could face future depleted supply and major challenges of the sustainability of this supply. Therefore, it is important to explore other venues to maintain the basin's long-term sustainability and meet future demand. Groundwater banking has become an integral component of any integrated water resources management plan to address both local and statewide water supply issues. Stakeholders from privately owned properties (or well owners) to local and state agencies are interested in the feasibility of groundwater banking in their respective basins.

Relative to the construction of surface water reservoirs, groundwater banking in subsurface aquifers is a less controversial, lower cost, and more environmentally benign approach. Groundwater banking has numerous economic and environmental advantages compared to surface water storage. It reduces losses to evaporation. It allows for greater regulation of natural inflows, without the need to construct of large surface reservoirs [5] . As with any water storage systems, the main purpose of groundwater banking is to store surface water from precipitation and rivers, when water is abundant, and utilize those groundwater sources when the surface water supply stocks is scarce. The major obstacles to groundwater banking are the water conveyance from the source to the area of injection, cost of construction and maintenance of injection wells and regulatory and permitting requirements. The current study assumes that groundwater recharge aspect is an acceptable water management technique.

METHODS

This study was prepared to evaluate the feasibility of groundwater banking in the Central Basin. A three dimensional (3-D) model of the Central Basin was developed using the GMS with MODFLOW (from here on referred as GMS) software package [6] . The model was simulated with three actual historical data scenarios: above normal, wet, and critical years. Once the model was developed and calibrated, it was utilized to examine the impact of groundwater banking. Additional groundwater was banked through injection wells for each scenario to study its feasibility in the Central Basin. In order to carry out groundwater banking analysis, the Central Basin data were collected and compiled into a model, which is described as follows.

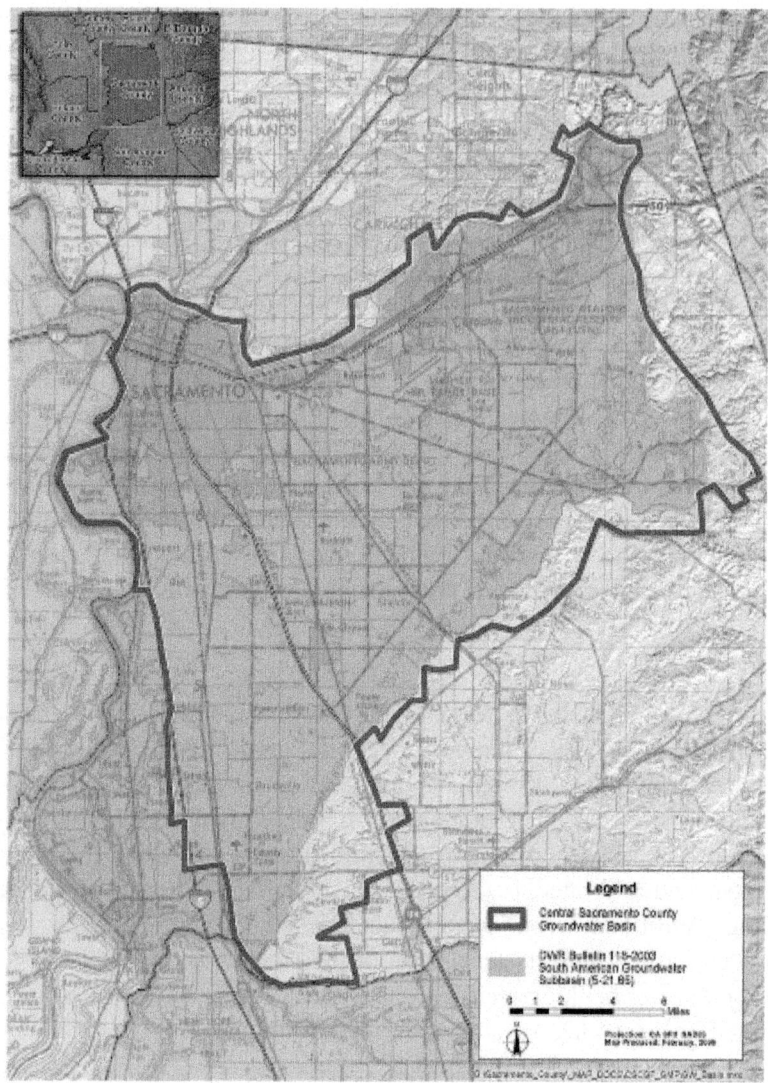

Figure 1: Study area.

Water Years Selection

The Sacramento Valley Water Year Type is determined by DWR's California Data Exchange Center's website (CDEC) based on Sacramento River and tributary runoff necessary to meet the Delta outflow criteria. Table 1 summarizes all possible water year types for the past seven years and the selected years of 2005, 2006, and 2008 [7].

Geology

Review of DWR publications was conducted to re-construct the complex geological formations in the Central Basin region. DWR publications provide detail information on layers of geologic formations in the region, but exact thicknesses of layers is not easily accessible from these publications. Therefore, the elevations of the different layers were estimated from a 2010 City of Roseville groundwater report called Sacramento Regional Model (SRM) [8] . In 2010, Aquaveo performed this groundwater study for City of Roseville to assist in the planning efforts for an aquifer storage and recovery. The SRM study included an area of approximately 1360 square miles in the Greater Sacramento Metropolitan Region which encompasses the area of this study (Central Basin). SRM used the information on the thickness of the post-Eocene continental deposits defined in the DWR Bulletin 118-6. Furthermore, SRM turned this geological information into a 3-D stratigraphic model of the region with estimated thicknesses for the layers of the geological formations. This study extracted those estimated thicknesses for the Central Basin region from the SRM. Table 2 lists ten layers of these geologic formations that were reproduced for the Central Basins model. The estimated geological formation thicknesses were entered into the GMS model as boreholes, which were then converted into a solid stratigraphy of the Central Basin.

Surface Elevations

The ground surface elevations are needed as an input parameter to the GMS model. The elevations were incorporated into the Central Basin region as the top of the surface soil for all the borehole data. The surface elevations were obtained from numerous sources including USGS and a number Geotechnical reports within the Central Basin region [9] -[13].

Precipitation

The precipitation data for the selected three water years 2005, 2006 and 2008 were retrieved from DWR's CDEC website. Table 3 shows the average annual precipitation for the selected years [7] . Average annual precipitation data for each of the selected water years were multiplied by the area (248,000 acres) of the Central Basin to obtain a volume of precipitation in acre-feet. Monthly variation of precipitation is not accounted for in this study. Table 4 summarizes the precipitation data and recharge rates [7].

River Stage

River stage elevations (annual average) along with the bottom elevations and river conductance are the key elements needed as river data into GMS. DWR's CDEC website provides stage elevation in feet for river stations, using the three letters station IDs. The river stage elevations used to develop the river system in Central Basin model are provided in Table 5 [7] . The locations of the river stations are shown in Figure 2 [7].

River Conductance

River conductance is related to the rate at which a unit of riverbed material can transmit fluids, and is used mainly in hydrology in relation to river and lake beds. It is an application of intrinsic permeability to a unit of material with a defined area and thickness. In hydrology, the magnitude of conductance affects the rate of groundwater recharge or interaction with groundwater. This parameter is used in computer modeling codes as GMS. Table 6 shows the initial River Conductance values used as starting point in Central Basin model.

Groundwater Usage

Groundwater use data is one of the key elements in developing the Central Basin GMS model. The SCGA's2007-2008 Basin Management Report provided a comprehensive groundwater use data, including the agricultural use of groundwater in the Central Basin (Table 7) [14] . The total groundwater and agricultural uses are annual averages; hence, monthly variation is not taken into consideration as part of this study.

Table 1: DWR water year type

Water year	Index	Year Type
	Million AF	
2004	7.51	Below Normal
2005	8.49	Above Normal
2006	13.2	Wet
2007	6.19	Dry
2008	5.16	Critical
2009	5.78	Dry
2010	7.05	Below Normal

Table 2: Ten geologic formation layers

Year	Groundwater Levels Predicted Contour Elevations from MSL (ft)		Groundwater Levels Predicted with 10,000 AF Contour Elevation from MSL (ft)	
	Max	Min	Max	Min
2006	160	−99	160	-92
2008	142	−123	142	-112

Table 3: Rain data for the selected year type

Water Year	Annual Average	Year Type
	inches	
2005	23.29	Above Normal
2006	25.36	Wet
2008	13.8	Critical

Table 4: Precipitation data.

Year	Average Annual Precipitation	Central Basin Area	Rain Volume	15% of Volume	15% of Volume	Recharge Rate
	(Inches)	(acre)	(AF/year)	(AF/year)	(ft^3/day)	(ft/day)
2005	23.29	248,000	481,327	72,199	8,616,951	0.0007977
2006	25.36	248,000	524,107	78,616	9,382,820	0.0008685
2008	13.8	248,000	285,200	42,780	5,105,793	0.0004726

Table 5: River stations & stage elevation

Station ID	Station Description	Lat.	Long.	Gage Elev.	Operating Agency	Stage Elev. 2005	Stage Elev. 2006	Stage Elev. 2008
AFO	American River At Fair Oaks	38.6	−121.272.0		USGS	99	105	91
MHB	Consumnes River at Michigan Bar	38.5	−121.0168.0		USGS & DWR	172	178	160
BEN	Mokelumne River near Thornton	38.3	−121.40.0		CA DWR	8	12	1
IST	Sacramento River at I Street	38.6	−121.527.0		CA DWR	12	16	2
SDC	Sacramento River above Delta Cross Channel	38.3	−121.510.0		USGS	5	9	−5

Table 6: Initial river conductance

River Name	River Cross-Sectional Area ft^2	b ft	K ft/day	C_b ft^2/day
Sacramento	3,000	20	0.1	15
American	2,000	20	0.1	10
Mokelumne	500	10	0.1	5
Consumes	200	5	0.01	0.4

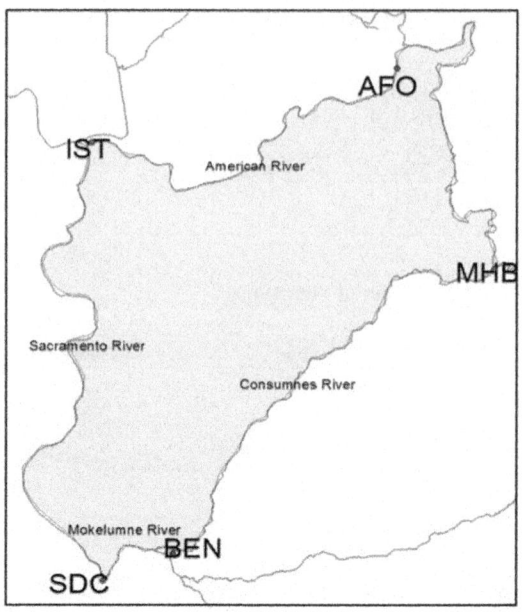

Figure 2: DWR's river station IDs & locations.

Model Development and Setup

The GMS model setup could be achieved using two approaches: the grid approach or the conceptual model approach [15] . The grid approach involves working directly with the 3-D grid, where conceptual model approach involves using GIS tools to construct the model and then it is converted into a 3-D grid. In the current study, a conceptual model approach was selected due to its applicability and flexibility to setup. The datum for the data needed for the GMS model was based on the State Plane Coordinate System with North American Datum 1983 (NAD83). A GIS shape-file containing rivers, streams and lakes in the Sacramento region was imported into GMS to provide information related to these water bodies. A background map was imported into the GMS model in order to provide ground surface elevations, visual guidance and spatial reference. The delineated project boundaries are shown in Figure 3.

Table 7: Groundwater usage

Year	Total Groundwater	Agriculture
	AF/year	
2005	244,026	167,062
2006	245,382	166,148
2008	247,067	164,320

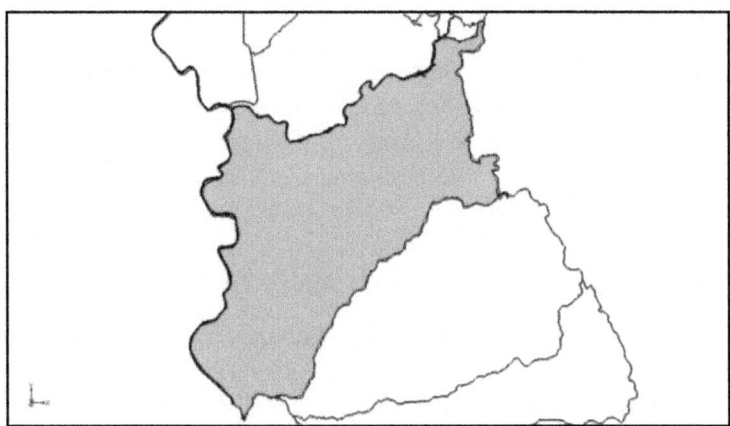

Figure 3: Central basin area & boundaries.

Boreholes for Sub-Surface Materials

Table 8: Well extraction flow rate

Year	Total Groundwater Use	Agricultural Original Use	Agricultural Use Revised	Final Total Agriculture Use		Well Extraction Rate
	(Acre-feet)			(Acre-feet)	(ft³/day)	(ft³/day)
2005	244,026	167,062	116,943	193,907	23,142,848	46,193
2006	245,382	166,148	116,304	195,538	23,337,413	46,582
2008	247,067	164,320	115,024	197,771	23,603,969	47,114

A total of twenty-two (22) boreholes were created and spread equally around Central Basin area to produce sufficient representative stratigraphy of the region. The borehole profiles are made of the geologic formation thicknesses estimated from City of Roseville study called SRM. Figure 4 shows the locations of all the boreholes across the Central Basin. Figure 5 shows the profile view of these boreholes.

Triangulated Irregular Network and Solids

Triangulated Irregular Network (TIN) method was used to calculate an intermediate ground surface elevations between points of known elevations by linear interpolation. Figure 6 shows the top soils with the interpolated ground surface elevations using TINs. Each of the 22 boreholes has specific elevations according to the estimated thickness of the geologic formation and ground surface elevation. TIN method uses those borehole elevations and connects them to create solids for different layers stratigraphy as shown in the Figure 7.

Pumping and Observation Wells

Five hundred and one (501) pumping wells were created in the conceptual model. The wells were classified based on the ownership entity: the DWR Wells (359), and the Non-DWR Wells (142).Figure 8 shows the locations of the DWR wells [7] , and Table 8 summarizes the wells extraction flow rate data [14].

THREE-DIMENSIONAL GRID AND MODFLOW MODEL

After entering all the center blocks (materials and their respective layers, rivers, wells, precipitation recharge, and boundary and initial conditions) of the Central Basin model, then model now has all the components for various simulations. The conceptual approach was then converted into 3-D Grid.Figure 9 shows the 200 × 200 × 14 grid containing the 3-D region of the Central Basin. Figure 10shows the model converted into the 3-D grids.

Model Calibration

DWR's spring of 2004 contour map was used as baseline for the model calibration, see Figure 11. The model elevations was run through manual process by deleting the 3-D grid, under which all MODFLOW simulations are stored and regenerating it for the next calibration.

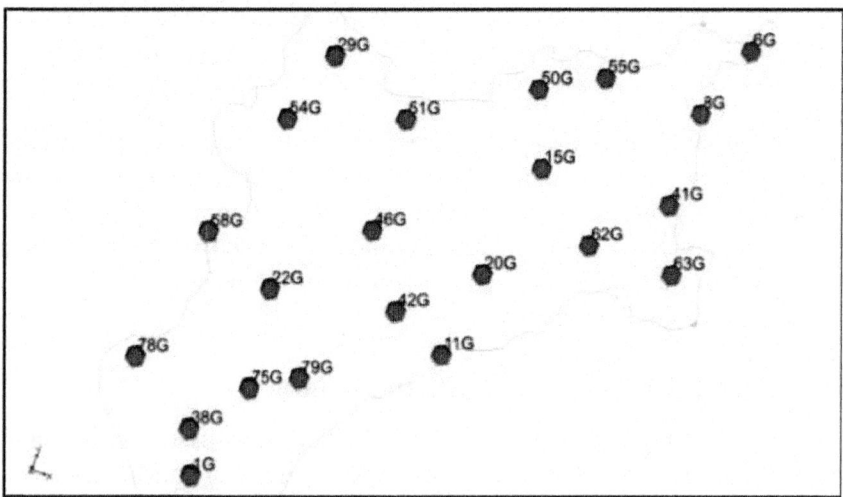

Figure 4: Borehole locations and labels.

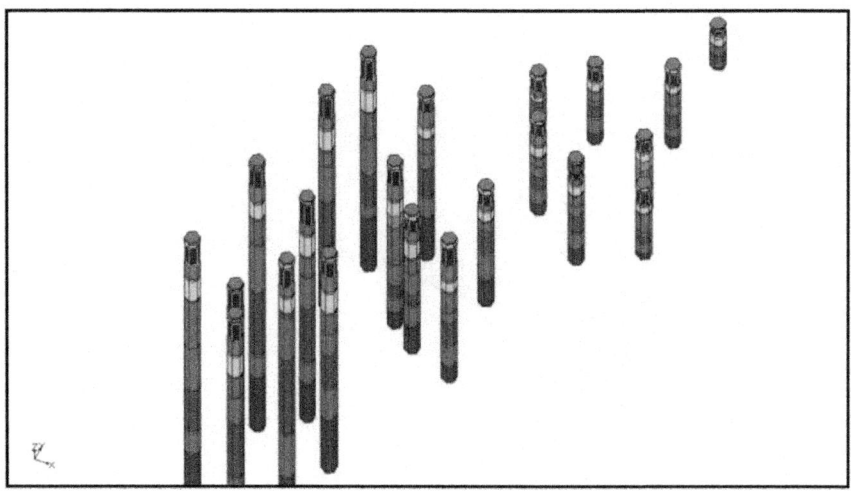

Figure 5: Profile of borehole.

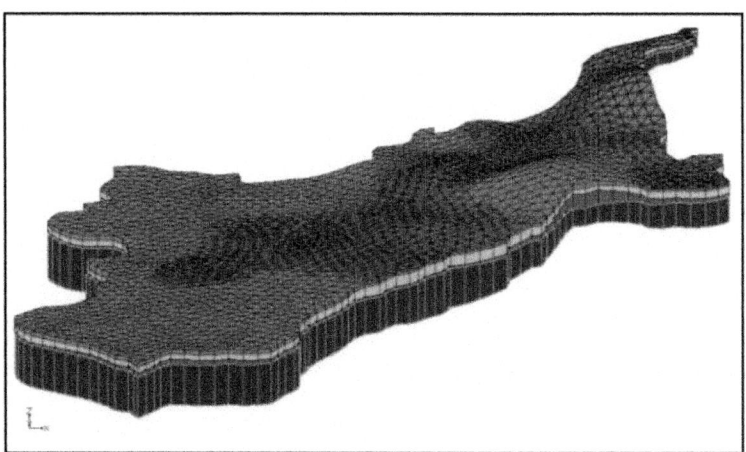

Figure 6: Top soil surface.

This process was repeated about two hundred (200) times in order to produce solutions that closely resemble the observed targeted contour map values as shown in Figure 11.

Parameter sensitivity analyses (approximately 200 runs) were performed and found that the model is more sensitive to the hydraulic conductivity (K value, feet/day) of the soil and geologic formations layers. The model was less sensitive to other elements such as river conductance, river stage elevations, porosity, specific yield and storage.

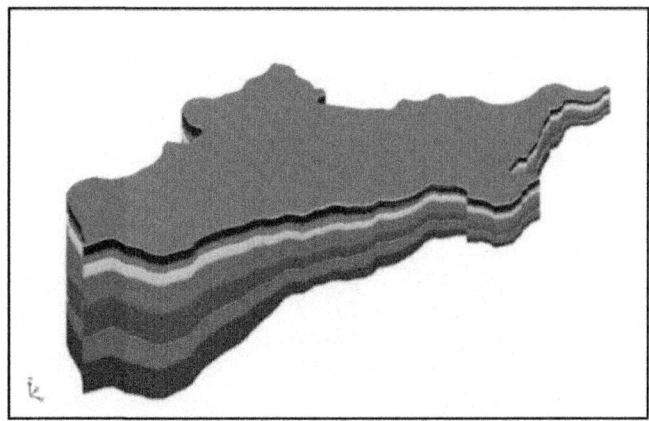

Figure 7: Solids stratigraphy.

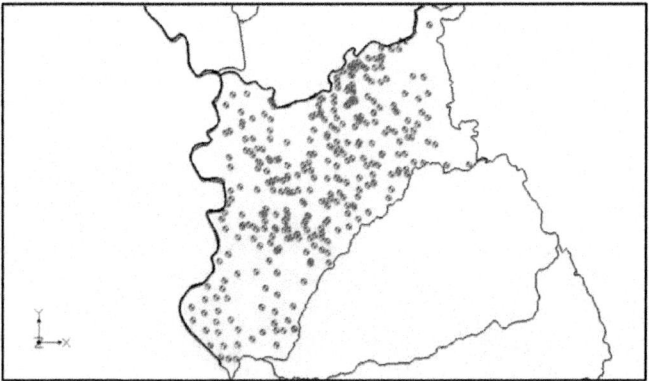

Figure 8: Locations of DWR wells.

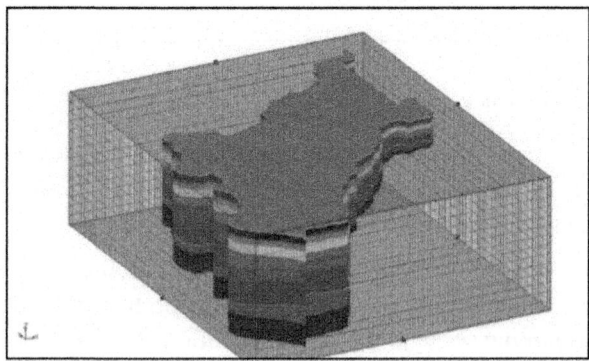

Figure 9: 3-D grid box.

Initial estimates of the hydraulic conductivities started at 0.1 - 10 ft/day and were adjusted until reasonable values 8 - 59 ft/day range were determined to match the existing conditions. The findings were in agree ment with a comprehensive research of the region's soil properties prepared for the Sacramento Area Flood Control Agency (SAFCA).

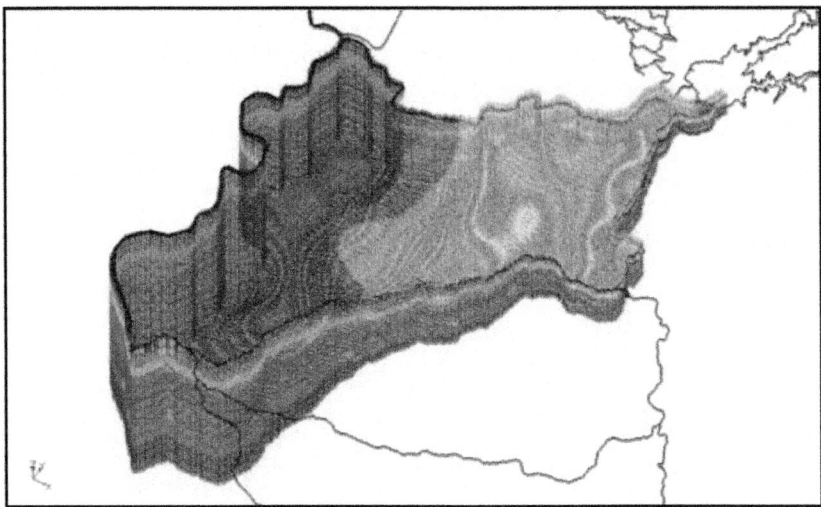

Figure 10: 3-D grid model of central basin.

The SAFCA study evaluated a broad range of reports and they concluded that the K values in the Sacramento area ranges from 5 - 139 ft/day [16].

Above Normal and Wet Years Calibration

The results of the 2005 and 2006 water years were very similar to the observed target (with a maximum difference of 10%), with the only exception of the location of the cone of depression.Figure 12 shows the model result for 2005 water year. In general, the formation of the cone of depression in all of the models was in agreement with historical records of SCGA and DWR. The results showed that the cone of depression of the historical record is shifted to the north when compare to the observed target cone of depression.

Critical Year Calibration

Results of the critical year type (2008) were also in agreement when compared to 2005 and 2006 water years' results. The cone of depressions was deeper, and the inflow from rivers and precipitation to the aquifer was smaller due to decrease in the amount precipitation and the river stage elevations.

Banking Wells

Banking wells were added to the Central Basin model at the location of the cone of depression. Sacramento Regional County Sanitation District (SRCSD), a regional wastewater treatment plant located in the city of Elk Grove, is currently delivering 15,000 AF of recycled water for agricultural use to local cities [17] . Therefore, the 10,000 acre-feet of water was selected as a starting goal of this study to bank. The groundwater banking wells for this study, were constructed to allow the use of any amount more than 10,000 AF if needed.

RESULTS

Critical Year with Banking

The process of injecting additional flow into the model as groundwater banking was successful and seems to produce conclusive results.

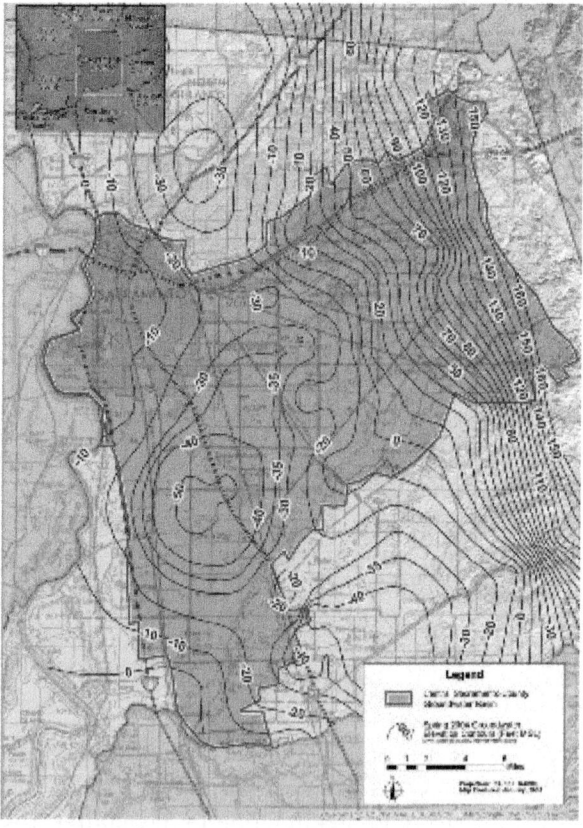

Figure 11: DWR contour calibration map.

The locations of the banking wells were chosen to be in the center of the cone of depression, where it was found to be the most effective locations for groundwater banking. The cone of depression for all three selected years (above normal, wet and critical) was in the same vicinity. The injection of 10,000 AF of groundwater in the effective area seems to have marginally altered the shape of the depression cone. There were no significant changes in the water table elevation elsewhere in the basin. Figure 13 shows the contour map of the basin with the injection of 10,000 AF for banking.

Wet Year with Banking

The results of the groundwater banking are summarized in Table9 It is clear that groundwater banking was more effective during the critical year since the water table elevation increased by 11 ft from 123 ft below Mean Sea Level (MSL) to 112 ft below MSL. While in the wet year (Figure 14), the change in groundwater table elevation was less dramatic. The change in groundwater table was only 7 ft from 99 ft below MSL to 92 ft below MSL.

CONCLUSIONS

This study presents the development, calibration, simulation and examination of the feasibility of groundwater banking for the Central Basin. The hydrologic analysis of the Central Basin was comprehensive and utilized gauge data that were available for rivers, precipitation and existing groundwater use as well as their interaction with the aquifer.

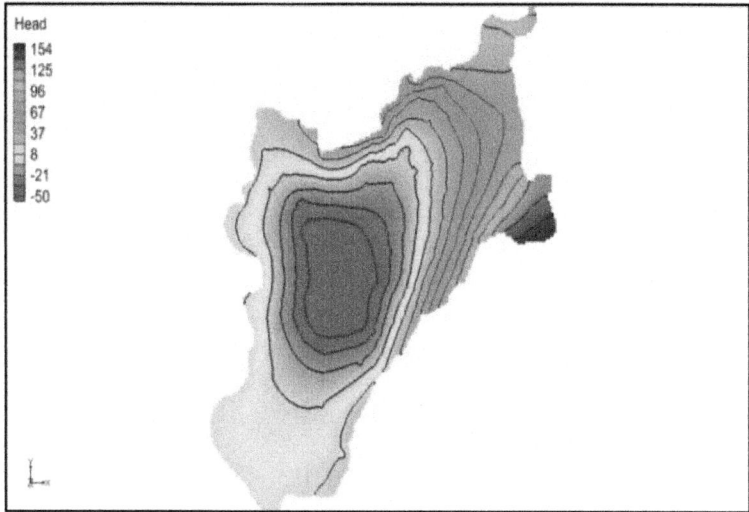

Figure 12: 2005 background contour map.

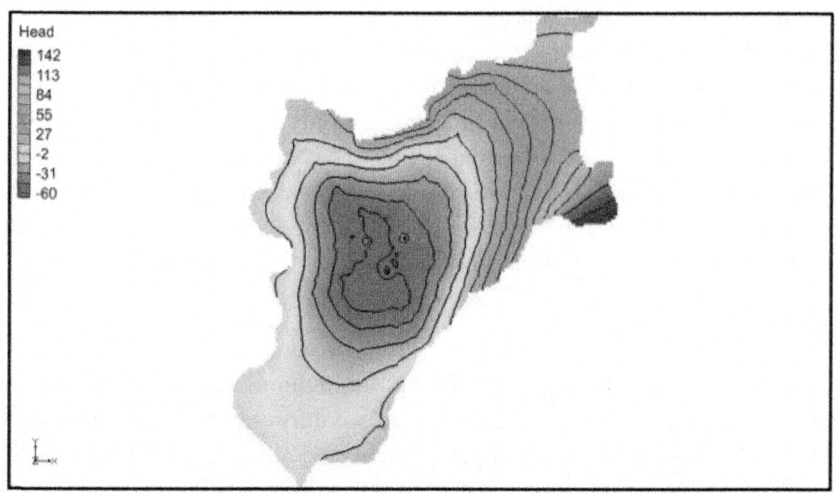

Figure 13: 2008 contour map with 10 K AF.

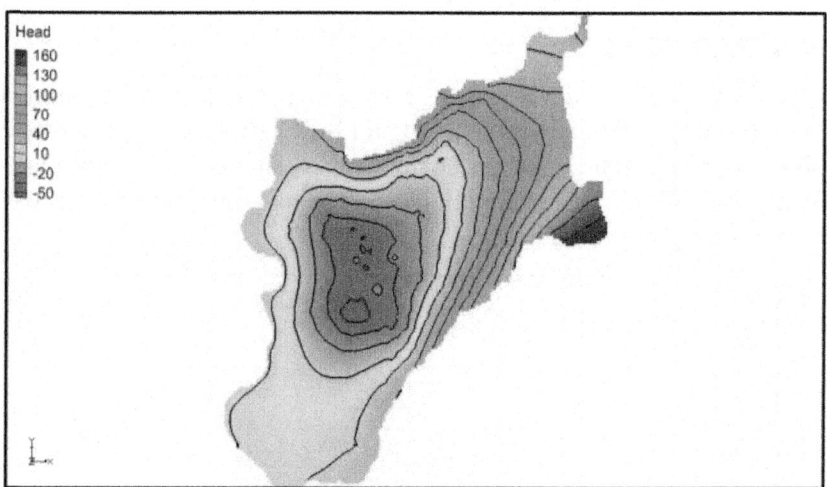

Figure 14: 2006 contour map with 10 K AF.

Table 9: Results of groundwater banking

Year	Groundwater Levels Predicted Contour Elevations from MSL (ft)		Groundwater Levels Predicted with 10,000 AF Contour Elevation from MSL (ft)	
	Max	Min	Max	Min
2006	160	−99	160	-92
2008	142	−123	142	-112

In addition, the GMS model was constructed with complex geologic formation, soil conditions of the region and all of the associated material properties. The model was constructed in a unique way that allows for future modification depending on data availability.

The results of this study showed that the cone of depression formulated during the different simulations is consistent with historical records. Historical analysis suggests that the formulation of the cone of depression is due to the excess of groundwater drafting in certain areas rather than the availability of resources for recharge. The size of the cone of depression was deeper during the critical year when compared to the above normal and wet years. It is further concluded that the banking up to 10,000 AF of groundwater during all three water year types (wet, above normal, and critical) is feasible in the Central Basin. The groundwater table elevation increased due to water banking in the all the scenarios studied. The highest increase in the predicted water table level was during the wet year scenario. The results also show that banking of 10,000 AF at the cone of depression will not cause negative impact elsewhere in Central Basin aquifer.

RECOMMENDATIONS

There were some limitations to this study, which could have influenced the result to a certain degree. First, the precipitation data used in the Central Basin model were annual averages; hence, monthly variations were not taken into consideration. Second, groundwater extraction rate was equally applied to all wells; hence, heavy users (i.e., Elk Grove Water District) were not taken into consideration. However, there are a number of recommendations that could be implemented in order to further improve this study. These recommendations are listed below:

- The extraction rate for each well is available and can be applied to individual wells. It will further increase the accuracy of the model.

- The computer that was used in this study to develop the Central Basin model that was not sufficient to refine the grid around the wells. It is recommended to use a computer with significantly larger storage and processing power to allow for more refinement of the GMS grid to improve accuracy.

- GMS is a strictly groundwater software and has limitation in applying precipitation and precipitation data. Therefore, if better surface runoff and percolation software is used and integrated with the GMS, it may improve the model further.

- Building an accurate stratigraphy of the region is one of the most complex tasks of groundwater model construction. This is subject to change according to the availability of new information from all sources including but not limited to federal, state, local governments and private property owners. Therefore, it is recommended to have significantly more man-hours to improve the stratigraphy of the Central Basin.

ACKNOWLEDGMENTS

The authors of this study would like to acknowledge the staff for help of the City of Roseville, California and for providing the data needed to complete the study. Furthermore, the authors would like to thank Aquaveo LLC for providing valuable information, resource and support towards this study and for donating the GMS software.

REFERENCES

1. Central Sacramento County Groundwater Management Plan (CSCGMP) (2006).

2. California DWR (2004) California's Groundwater: South American Groundwater Sub-Basins Number: 5-21.65. Bulletin 118-6.

3. California Department of Water Resource (DWR) (1974) Evaluation of Ground Water Resources: Sacramento County. Bulletin 118-3.

4. DWR (1978) Evaluation of Ground Water Resources: Sacramento Valley. Bulletin 118-6.

5. Natural Heritage Institute (NHI) (1998) Feasibility Study of a Maximal Program of Groundwater Banking.

6. Groundwater Modeling System (GMS) with MODFLOW software. Version 8.0. (2011) Aquaveo, LLC, Provo, Utah.

7. California DWR's California Data Exchange Center Website. http://cdec. water.ca.gov/

8. City of Roseville (2010) Sacramento Regional Model Groundwater Modeling Report. Aquaveo LLC.

9. Morris, D.A. and Johnson, A.I. (1967) Summary of Hydrologic and Physical Properties of Rock and Soil Materials, as Analyzed by the Hydrologic Laboratory of the U.S. Geological Survey, 1948-1960. USGS Water Supply Paper: 1839-D.

10. Folsom Cordova Unified School District (2008) Geologic Hazard and Geotechnical Engineering Report. Wallace Kuhl & Associate Inc.

11. Youngdahl Consultating Group Inc. (2007) Geotechnical Engineering Study—Update 2007. Seasons at Laguna Ridge.

12. CSA Water Pipeline and Florin Road Sewer (2009) Geotechnical Baseline Report. ENGEO Inc.

13. Elk Grove Promenade Center (2006) Geotechnical Engineering Report. Wallace Kuhl & Associate Inc.

14. Sacramento Central Groundwater Authority (SCGA) (2008) Basin Management Report 2007-2008.

15. Aquaveo LLC (2011) GMS 8.0 Tutorials. Retrieved from Aquaveo Website.http://www.aquaveo.com/gms-learning

16. Sacramento Area Flood Control Agency (SAFCA) (2009) Appendix A: Groundwater Impact Analysis. Luhdorff & Scalmanini Consulting Engineers.

17. Sacramento County Regional Sanitation District (SRCSD) (2011) South Sacramento County Agriculture & Habitat Lands Recycled Water Project Programmatic Feasibility Study.

Chapter 4

GROUNDWATER SOLUTION TECHNIQUES: ENVIRONMENTAL APPLICATIONS

Sarva Mangala PRAVEENA[1], Mohd Harun ABDULLAH[1], Ahmad Zaharin ARIS[2], Kawi BIDIN[1]

[1]School of Science and Technology, Universiti Malaysia Sabah, Kota Kinabalu, Sabah, Malaysia.

[2]Department of Environmental Sciences, Universiti Putra Malaysia, Selangor, Malaysia

ABSTRACT

Groundwater models provide a scientific tool for various groundwater studies which include groundwater flow, solute transport, heat transport and deformation. However, without a good understanding of a model, modeling studies are not well designed or the model does not represent the natural system which being modeled long term effects may results. Thus, this review has focused and reviewed the types of solution techniques in terms of advantages and limitations. The findings are vital to improve the model conceptualization and understanding of the uncertainty in model results. On the same hand, it acts as guide and reference to groundwater modeler, reduces the time spent in understanding the solution technique and complexity of groundwater models, as well as focus ways to address the groundwater problems and deliver modeling output more efficiently.

INTRODUCTION

According to [1], groundwater modeling covers different aspects of the system behavior. Groundwater modeling studies have four potential relevance processes which include groundwater flow, solute transport, heat transport and deformation. According [2,3], groundwater modeling has turn out to be a crucial tool in decision making and planning in environmental management. Decision making and planning processes in environmental management are associated with water resource allocation, complex development and requiring multidisciplinary information for evaluating their effects on a social, economic and environmental level [4]. Generally, most of the groundwater modeling studies are conducted using either deterministic models, based on precise description of cause-and-effect or stochastic models based on the probabilistic nature of a groundwater system [5,6]. The main components of groundwater modeling are selecting the natural system which the model is designed, creating the conceptual representing the natural system, models representing the controlling mechanism, solution of the model, calibration and validation of the model along with simulation [7,8].

There are enormous amount of groundwater models to study the cause and effect or the probabilistic nature of a groundwater system. It is an ad-vantage to classify them in groups based on criterias such as aquifer type, techniques used, type of aquifer simulated and the dimension of the problem [9]. [10] stated that the classification of groundwater models can be done based on model objectives, processed modeled, physical system characteristics modeled and mathematical approaches. According to International Ground Water Modeling Center (IGWMC), there are many various manners in groundwater models classifications (flow, media, transport, temperature, phases, chemical reaction, dispersion, thermodynamics, fractured rock, vapor transport, variable saturated, saturated) that a specific and systematic classification cannot be developed. A detailed explanation of these classifications can be found in [10].

Various solution techniques are a crucial component in groundwater models [6]. Solution techniques in groundwater modeling activities are to follow a multilevel approach. Multi-level approach involves data collection of groundwater flow and mass, contaminant transport and advection-dispersion equations, evaluation of the data and final decision to select the model. An understanding of various solution techniques is vital due to complexity in groundwater modeling and universal importance perspective. Era of numerous groundwater models development has been stimulated by high advance of computer technology and programming techniques. Yet the current numerous model development and groundwater complexity often leave those involve in groundwater studies spend a lot of time in understanding the solution techniques. This increased time resulted in less time spent in understanding the system. Thus, there are many gaps in our understanding of groundwater modeling which limits our capacity. Various groundwater models development have exposed with many reviews on the favors and disfavors of these models [6,8,11]. However, there are limited reviews on the solution techniques of these groundwater models although they are crucial components utilized in groundwater modeling. While a number of these solution techniques are focused on the types of models and applications in real world [11–14], a lack of quantitative information on the advantages and limitations of these tools impedes the use of these tools for real-world applications.

An understanding of various solution techniques is crucial due to complexity in groundwater modeling. This work was intended primarily as a guide and reference for the practitioner who is trying to simulate groundwater in their site of interest. This attempt is a way to lessen the time spent in understanding the solution technique and complexity of groundwater models, as well as focus ways to address the groundwater problems to render modeling output more effectively. The conceptual framework of the review was based on the types of solution techniques available in groundwater studies. An assessment of mutual understanding, advantages and limitations of all the solution techniques is applied to all kind of groundwater modeling studies and not limited to any particular purpose or equations. It is an attempt to reduce the time spent in understanding the solution technique and complexity of groundwater models and represent focus ways to address the groundwater problems and render modeling output more effectively.

VARIOUS SOLUTION TECHNIQUES ASSESSMENT

According to [8], the term model has different meanings. Combinations of all model components are suitable for groundwater model. However, term model is also used in a part of various solution techniques. Thus, the term model will also be used in a part of solution technique in this review. Numerous sophisticated solution techniques or model are currently available to overweigh the accuracy of the groundwater system representation [15]. The groundwater solution techniques comprise from simple to complex [6]. According to [2] until early 1970s, physical and analog models were widely used as mathematical models solving groundwater problems. As groundwater modeling techniques boosted with extensive computer programmings, various solution techniques have been developed to solve the systems of mathematical equations. The simplest classification was done by [12] and [14], where the solution techniques are divided into two broad groups namely physical models and mathematical or numerical models. Solution techniques grouping done by [11] listed that groundwater models are divided into four broad groups which are porous media, analog, electric analog and digital models. Along with the advent of computers, groundwater modeling has focused on the numerical models expressing the groundwater flow and transport studies. However, these models (analytical, physical, analog, porous, empirical and mass balance) are still needed to investigate and validate new models. The requirements are to examine and analyze whether certain assumptions underlie the new models are valid. The conceptual framework of this review was based on the types of solution techniques listed by [8] as showed in Figure 1.

SOLUTION TECHNIQUES EVALUATIONS

It is very important to have strong understanding with a model in order to know the advantages and limitations of each solution techniques. Perspectives of advantages and limitations of the solution techniques were evaluated in this review.

Analytical Models

Analytical models are the rapid way to analyze physical characteristics and conceptual behavior of groundwater system compare to other models. This is because it uses an exact analytical solution for specific field applications. On the other hand, analytical models are only limited to steady and uniform groundwater problems involving small parts of study area and bulky to transport problems. Table 1 presents other points of advantages and limitations of analytical models.

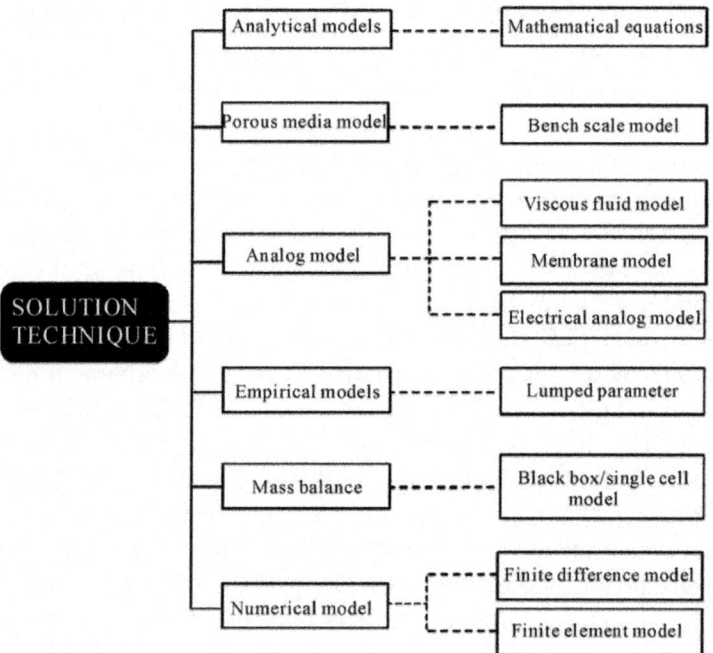

Figure 1: Types of groundwater models.

Porous Media Models

Porous media or bench-scale models belong to the group of hydraulic models which has been widely used in hydraulic engineering. Porous media models are suitable to use at any dimensionality, any type of groundwater flow and transport problems (variable saturated, heterogeneity, anisotropy, phreatic, steady, unsteady, advection, dispersion, sorption, decay and reactions). Information about porous media is presented in Table 2.

Analog Models

In terms of demonstration and education tools, analog models are still widely used for groundwater studies. Analog models (viscous fluid, membrane and lumped models) are not suitable for groundwater transport. The models have limited capability to involve with advection, dispersion, sorption, decay and reactions studies in groundwater. The views on advantages and disadvantages of analog models are detailed in Table 3.

Table 1: Applicability of analytical models

Model type	Analytical model
Advantages	• Simple [6,16] • Economical/ inexpensive [2,3,6] • Rapid way to analyze physical characteristics of groundwater [2,3,20] • More efficient than other models [6,9,16] • Can form useful complements to any numerical models [25,26] • Can used either for verification or being part of numerical models [16,17] • An important and useful tool for estimating fate and transport parameters from field or laboratory data [16,17] • Provide more insight into conceptual behavior of the groundwater system [3] • Does not introduce errors due to the numerical diffusion and approximation by the finite difference model [12]
Limitations	• An exact analytical solution may outweighed by the errors introduced by simplifying assumptions of complex field environment [9,10] • Complex and cumbersome in transport problems [2] • Limited to cases with steady and uniform flow problems [2] • Relatively simple initial and assumptions in boundary conditions. Hydrogeological boundary conditions must be idealized to fit the model [2] • Professional judgment and experience in field application are needed to apply the analytical model [2] • Suitable to solve groundwater problems involving small parts of aquifer systems or small area extent [9,18] • Could not handle spatial/temporal variations in groundwater system [18]

Empirical Models

Empirical models are useful to use when detailed site specific data are lacking or impractical situation to simulate fine-scale processes. Lack of understanding in the processes involve in study area, these models can be misused or misunderstood as the models are easy to employ as well as lumping process together will mask the disadvantages of these models. Table 4 summarizes the information on empirical models.

Table 2: Applicability of porous media models

Model type	Porous media model
Advantages	• Relatively straightforward and simple [19, 20] • Allow the study of special aspects of groundwater flow and transport under almost natural condition [19,20] • Useful to enhance site characterization and features [9] • Good demonstration and education tools for students [4,7,20] • Obeys laws that govern other physical systems including laminar flow of fluids and heat [4,6,7] • Good starting point for groundwater modeling beginners [4]
Limitations	• Capillary rise takes place in such models is far larger than that which actually occurs in a real field situation [13] • Difficult to visual and identify the water table [7,13] • Time consuming and prohibitively costly [5]

Table 3: Applicability of analog models

Model type	Analog model
Advantages	• Illustrative and still widely used for demonstration purposes of groundwater flow [4,21] • Versatility and can readily study a variety of aquifer conditions [8] • True for groundwater flow without natural recharge if the weight of the membrane is small [4] • Inexpensive tools to use to visualize groundwater stress [4] • Useful tool to help the inexperienced earth scientist to understand about groundwater hydraulics [4] • Solves problems concerning the phreatic surface for transient and steady flow conditions[4,7,21]
Limitations	• A good care is required in the model construction because flow rate varies with the cube width [4,7] • Temperature is also another factor need to be focused [4,5] • Limitation on applications involving nonlinear conditions of varying transmissivity in unconfined aquifers and two-fluid flow problems [7,13] • Also limited applications in groundwater lowering in construction field [21] • Electric potential is unaffected by gravity, therefore it requires adjustments [22]

Mass Balance Models

Mass balance model is also known as the black box or single-cell model. It is also a numerical model in its simplest form. In mass balance models, the averaging of an entire area is a crude approximation. Evaluation of field data is only involves in and out fluxes. Table 5 details the information of mass balance models.

Table 4: Applicability of empirical model

Model type	Empirical model
Advantages	• Impact the accuracy of the model predictions [23,24] • Suitable to use when detailed site specific data are lacking and appropriate when it is impractical to simulate fine-scale processes [4] • Representing an entire groundwater problem employs a series of physical laws, empirical laws and conservative assumptions to represent the problem of interest [1,4,23,24] • A good alternative method [23,24] • Provide useful predictions without the costly calibration time [23,24]
Limitations	• Lack of understanding of process involved and only a temporary solution to assist analysis [7, 24] • Can be misused and misunderstood because they are easy to employ [4] • Lumping processes together will mask the limitations of these models [7]

Table 5: Applicability of mass balance model

Model type	Mass balance model
Advantages	• The simplest form of numerical model. The best fitted in numerical modeling [4,14] • Very useful which leads to an examination of the global mass balance [14] • Easy to use [4,14] • Efficiently aid in the analysis of the impact of the management options [14] • Suitable to use when detailed site-specific data are lacking or impractical situation to simulate fine-scale processes [14] • An important part in more complexes of numerical models [8]
Limitations	• Lack of understanding of the processes involved [4] • Acts as a temporary solution to aid analysis [4,14] • Can be misused or misunderstood because they are easy to use [25] • Applicable only in limited circumstances and masked by lumping process together [10,14]

Numerical Models

Among of the solution techniques assessment, numerical models were found to have more advantages over other solution techniques. They are such as it solves both simple and complex groundwater problems, capable to used almost of any type of groundwater system and impose no restrictions on the initial conditions, boundary types as well as characteristics of the groundwater. The most advantage in numerical models is that the models utilize the latest advances in computer technology without writing any computer codes. Numerical models which employ the latest computer technology also have limitations in terms of accuracy, errors and codes. Accuracy of numerical output mainly depends on the availability of soil hydraulic information, errors in numerical dispersion are hard to be identified as well as special codes are need for specific groundwater problem (Table 6).

Table 6: Applicability of numerical model

Model type	Numerical model
Advantages	• Employed with the latest and recent advances in computer technology [4,5,11,13] • Solves both simple and complex groundwater problems [4,7,13,26,27] • Dominated the complex study of groundwater problems as it solves both simple and complex one, two or three dimensional problems [4,7,13, 15] • Capable to simulate almost any type of groundwater situation [5,7,17] • Well suited to exploring hypothetical scenarios [15,27] • Can easily handle spatial or temporal variations of groundwater system [6,11] • Impose no restrictions on the initial conditions, boundary types, characteristics of the groundwater or investigated solute [5,10] • Computer programs for most groundwater problems are available easily and the users can apply relevant computer programs without writing any computer code [4,7,13,26,27]
Limitations	• Time consuming for data collection and input [4,7,11,13] • Require much information to characterize the system [28] • Expensive models [28] • Special codes are required for specific problems, such as density-dependent flow and coupled saturated unsaturated flow [15,27] • Accuracy of the results of numerical models mainly depends on the availability of information about the hydraulic properties of the subsoil [28] • Errors in numerical dispersion [28] • Uncertainty of the model predictions is hard to quantify [28]

CONCLUSIONS

This review has focused and reviewed the types of solution techniques available in groundwater modeling studies. Assessment of six solution techniques namely analytical, porous media, analog, empirical, mass balance and numerical models was done to give a clear understanding of each solution techniques. Advantages and limitations of all the solution techniques were listed and analyzed. Analytical, porous media and mass balance models are simple and appropriate to use in groundwater modeling studies. In terms of demonstration and education tools, porous media and analog models are still widely used for groundwater studies. Empirical and mass balance models are useful to use when detailed site specific data are lacking or impractical situation to simulate fine-scale processes. The most benefit of numerical models is it utilizes the latest advances in computer technology without writing any computer codes as well as solves both simple and complex of any groundwater problems. On the other hand, limitations of analytical models are only limited to steady and uniform groundwater problem involving small parts of study area. Porous media and numerical models face time consuming for data collection and expensive as their constraints in the applications. Empirical and mass balance models face lack of understanding in the processes involve in study area and can be misused or misunderstood. In the view of analog models, they are not suitable for groundwater transport. Moreover, errors in numerical dispersion are hard to be identified as well as special codes are need for specific groundwater problems. As a final note, it is important to point out that a good understanding of various solution techniques act as guide and reference to groundwater modeler. Besides, it reduces the time spent in understanding the solution technique and complexity of groundwater models, as well as focus ways to address the groundwater problems and render modeling output more effectively.

ACKNOWLEDGEMENT

The first author gratefully acknowledges the support by National Science Fellowship (NSF) Scholarship under sponsorship of Ministry of Science, Technology and Innovation (MOSTI), Malaysia for her doctoral study. Sincere appreciation is also extended to the reviewers for their helpful comments and suggestions which have improved the quality of this paper.

REFERENCES

1.   L. W. Canter, D. M. Fairchild, and R. C. Knox, "Ground water quality protection," CRC Press, Boca Raton, Florida, 1988.

2. J. Bear, M. S. Beljin, and R. R. Ross, "Fundamentals of groundwater modeling," United States Environmental Protection Agency, 1992.

3. P. K. M. van der Heijde, "Quality assurance in computer simulations of groundwater contamination," Environmental Software, Vol. 2, pp. 19–25, 1987.

4. E. Manoli, P. Katsiardi, G. Arampatzis, and D. Assimacopoulos, "Comprehensive Water Management Scenarios for Strategic Planning," Global NEST Journal, Vol. 7, pp. 369–378, 2005.

5. S. M. Praveena, M. H. Abdullah, A. Z. Aris, and L.C. Yik, "A brush up on seawater intrusion models," in the Proceeding of Third Regional Symposium on Environment and Natural Resources, Kuala Lumpur, pp. 313–324, 2008.

6. C. P. Kumar, "Pitfalls and sensitivities in groundwater modeling," Civil Engineering, Vol. 84, pp. 116–120, 2003.

7. V. S. Singh and C. P. Gupta, "Groundwater in a coral island," Environmental Geology, Vol. 37, pp. 72–77, 1999.

8. K. Spitz and J. Moreno, "A practical guide to groundwater and solute transport modeling," John Wiley and Sons, New York, 1996.

9. M. E. Thangarajan, "Resource evaluation, augmentation, contamination, restoration, modeling and management," Capital Publishing Company, 2007.

10. J. R. Boulding and J. S. Ginn, "Practical handbook of soil, vadose zone, and ground-water contamination: Assessment, prevention and remediation," Lewis Publishers, Boca Raton, Florida, 2004.

11. D. K. Todd, "Groundwater hydrology," Second Edition. John Wiley & Sons, New York, 1980.

12. M. P. Anderson and W. W. Woessner, "Applied groundwater modeling: Simulation of flow and advective transport," Academic Press, Inc., San Diego, 2002.

13. N. Krešić, "Hydrogeology and groundwater modeling," CRC Press, Boca Raton, Florida, 2006.

14. W. C. Walton, "Groundwater resource evaluation," McGraw-Hill Education, 1976.

15. K. McGillicuddy and T. Sovich, "Strategies for operation of orange county water district Talbert seawater intrusion barrier, California," ASCE, New York, 1996.

16. A. M. M. Elfeki, G. J. M. Uffink, and F. B. J. Barends, "Groundwater contaminant transport: Impact of heterogeneous characterization: A new view on dispersion," Taylor & Francis, 1997.

17. N. Emekli, N. Karahanoglu, H. Yazicigil, and V. Doyuran. "Numerical simulation of saltwater intrusion in a groundwater Basin," Water Environment Research, Vol. 68, pp. 855–866, 1996.

18. L. F. Konikow and T. E. Reilly, "Groundwater modeling," In: The handbook of groundwater engineering, CRC Press, Boca Raton, Florida, 1995.

19. L. Konikow and J. Mercer, "Groundwater flow and transport modelling," Journal of Hydrology, Vol. 100, pp. 379–409, 1988.

20. J. J. Fried, "Groundwater pollution: Theory, methodology, modelling, and practical rules," Elsevier Scientific Publishing Company; Amsterdam-Oxford-New York, 1975.

21. R. Bowen, "Groundwater," Springer; London, 1986.

22. J. Wainwright and M. Mulligan, "Environmental modeling: finding simplicity in complexity," John Wiley and Sons, New York, 2004.

23. K. R. Rushton, "Groundwater hydrology: Conceptual and computational models," John Wiley & Sons, New York, 2003.

24. Environmental Protection Agency, "Models and computers in ground-water investigations," 1991, http://www. cepis.ops-oms.org/muwww/fulltext/repind46/models/models.html.

25. V. Batu, "Applied flow and solute transport modeling in aquifers: fundamental principles and analytical and numerical methods," CRC Press, Boca Raton, Florida, 2006.

26. G. B. Maxey, W. Back, and D. A. Stephenson "Contemporary hydrogeology," The George Burke Maxey memorial volume, Elsevier, 1979.

27. M. Kasenow, "Determination of hydraulic conductivity from grain size analysis," Water Resources Publication, 2002.

28. E. Holzbecher, "Modelling density-driven flow in porous media," Springer Publisher, Heidelberg, 1998.

Chapter 5

CONTRIBUTION OF THE SENSITIVITY ANALYSIS IN GROUNDWATER VULNERABILITY ASSESSING USING THE DRASTIC METHOD: APPLICATION TO GROUNDWATER IN DABOU REGION (SOUTHERN OF CÔTE D'IVOIRE)

Jacques Édoukou Djémin, Jean Kan Kouamé, Kouakou Serge Deh, Amani Tawa Abinan, Jean Patrice Jourda

Department of Sciences and Water Technology and Environment Engineering (Laboratory of Remote Sensing and Spatial Analysis Applied to Hydrogeology), Félix Houphouët-Boigny University, Abidjan, Côte d'Ivoire

ABSTRACT

The groundwater constitutes the main source of drinking water for the populations in the Dabou region which is marked by a multiplication of socio-economic activities. The quality of groundwater is increasingly tested by diverse sources of pollution caused by these human activities. In order to preserve their quality against any form of contamination, the present study aims to assess the groundwater vulnerability in this region and to highlight the relative importance of hydrogeolo- gical parameters which will be taken into account in this assessment. The assessment of the intrinsic vulnerability is to identify the most sensitive zones in order to prevent the groundwater pollution risks on the surface of the ground. To do it, the DRASTIC method is applied through a GIS. The GIS has also used to perform sensitivity analysis through the map removal and the single- parameter sensitivity analysis tests. The indexes calculated for the DRASTIC vulnerability map vary from 95 to 187 of the North towards the South. This vulnerability map presents four classes: very high (26.22%) in the South and the East, high (37.71%) in the Center, the North-East and the North-West, moderate (34.73%) to the North and the West and low (1.34%) in the North. The DRASTIC vulnerability map is heavily influenced by the impact of vadose zone and the depth to water table according to the first test. For the second test, it is the impact of vadose zone, the aquifer media and the soil media which have a more significant impact on the vulnerability map. Both sensitivity analysis tests confirm that the impact of vadose zone therefore sediment type is more implied in this assessment of the groundwater vulnerability in the Dabou region.

INTRODUCTION

The groundwater is the main source of drinking water for some populations. The estimates indicate that nearly two billion people in the world depend directly on the groundwater [1] . Groundwater used not only for human consumption but also the development of socio-economic activities. However, they are exposed to many sources of diffuse and punctual pollution resulting from human activities, particularly the agricultural practices [2] . The progress and the generalization of industrialization have also increased the risk of groundwater pollution by using of hazardous substances [3] . The introduction of pollutants into water tables deteriorates the quality of groundwater and reduces their consumable nature [4] . Given this situation, it is necessary to study the aquifers vulnerability, which is defined as the set of natural geological and hydrogeological characteristics that determine susceptibility of groundwater to contamination by human activities [5] . The vulnerability assessment therefore aims to prevent potential risks of pollution by the delimitation of sensitive areas in order to control the activities at the ground surface to protect the groundwater quality. The groundwater of the study area is exploited in both urban and rural areas to satisfy the water requirements of the populations. However, in some cases, nitrate concentrations of exploited water can reach 51.7 $mg \cdot L^{-1}$ [6] . This high concentration of nitrates in collected water can be related to several factors such as the intensive use of agricultural inputs and pesticides in agro-industrial plantations for the improvement of the productions. Moreover, the uncontrolled urbanization favoring the installation of inadequate systems of cleansing and the insufficiency of these systems for a better management of the domestic and industrial waste water rejections can be the causes of a real deterioration of these waters quality. Added to this is the storage of household waste promoting the production of leachate (lixiviates) which can cause the groundwater contamination in this region. Indeed, to limit and prevent contamination risks of these waters in order to preserve their quality, it is necessary to establish the vulnerability map which is under the influence of several hydrogeological parameters. This is firstly to assess the groundwater vulnerability from the DRASTIC method that uses seven hydrogeological parameters and secondly to assess the relative importance of each parameter based on sensitivity analysis tests. The sensitivity analysis will make it possible to apprehend the most determining parameters in the assessment of the intrinsic vulnerability in this region. Application of DRASTIC method and sensitivity analysis tests requires the use of a Geographic Information System (GIS) which is the gateway to the database and output decisional thematic maps.

PRESENTATION OF STUDY AREA

The study area is the Dabou region which is a coastal area of southern Côte d'Ivoire (Figure 1). It is between longitudes 4°10' and 5°00' West and latitudes 5°00' et 5°40' North with an estimated area at 2435 km². The population is estimated at 238,294 inhabitants according to the last population census [7] . The study area consists of three departments: Dabou which occupies much of the area, Jacqueville and Grand Lahou. The climate is equatorial transition characterized by four seasons with two dry seasons which alternates with two rainy seasons. The average rainfall ranges around 1400 mm. The direct infiltration feeding the water tables is estimated on average at 382 mm a year which represents 26.30% of the rainfall. This coastal area is characterized by a hydrographic network with an average drainage density equal to 1.7 km.km⁻² dominated by Ira River and Ebrie Lagoon. The Agneby and Bandama rivers respectively to the East and West of the area are hydraulic limits of the basin.

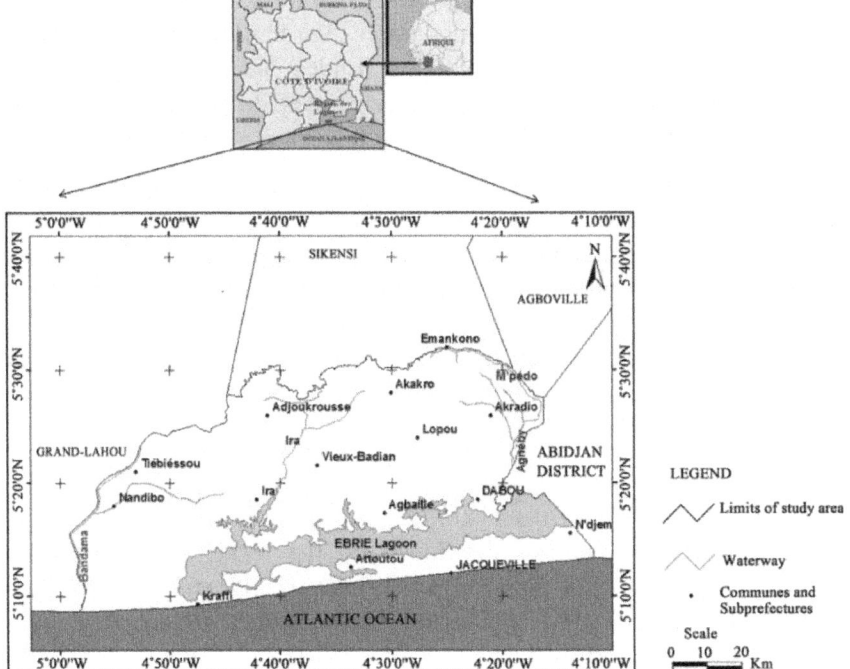

Figure 1: Location map of the study area.

Geological and hydrogeological, the study area belongs to the coastal sedimentary basin composed of post- Eburnean formations characterized by quaternary clayey sands and vases and Continental Terminal clayey sands [8] . This cretaceous to quaternary basin has enormous potential groundwater [9] . These groundwater resources are contained in aquifers three levels of unequal importance [10] . The aquifer system contains the quaternary water table and that of the continental terminal. The aquifer system is exploited by more than 217 boreholes (drillings) with an average speed of 18.79 $m^3 \cdot s^{-1}$.

MATERIALS AND METHOD

Data Acquisition and Materials

The drillings data used in this study were collected from the Water Resources Direction. The drillings data such as transmissivity which were used to estimate the hydraulic conductivity values starting from the thicknesses of calculated water table were collected in previous study. Identification of geological formations that make up the aquifer system in the study area required the drilling logs that exist in previous research of [6] . The geological map (scale: 1/200,000) provided by the Cartography and Remote sending Center has delineated the study area. The Digital Elevation Model (DEM) used in this study is a SRTM (Shuttle Radar Topography Mission) image. It was collected in 2008 by the American Spatial Shuttle "Endeavour" available from (http:// strm.csi.cgiar.org) and was used to delineate the DEM of the study area which enabled subsequently estimating slopes of this zone.

The GIS software used in this study is ArcGIS 9.3 offered in three levels of licenses (ArcView, ArcEditor, and ArcInfo). The three modules used in this study are: ArcCatalog that enables the management of the database or the geodatabase; ArcMap used for geoprocessing and ArcScene used for overlaying the information planes. ArcGIS 9.3 has several ex-tensions are those used "Spatial Analyst or GRID and 3D Analyst". Excel software has allowed the entry of alphanumeric data that is imported into the ArcGIS environment through its module ArcCatalog.

DRASTIC Method

The DRASTIC method was developed by the services of the American Agency of Environmental Protection USEPA [11] . It estimates the potential for pollution and to assess the vertical groundwater vulnerability [12] . It takes into account most of the hydrogeological factors that affect and control the flow of groundwater [13] . It is a method to index weighting with seven parameters whose initials form the acronym DRASTIC: Depth to water table

(D), net aquifer Recharge (R), Aquifer media (A), Soil media (S), Topography (T), Impact of vadose zone (I) and Finally the hydraulic Conductivity of the aquifer (C). The development of the intrinsic vulnerability map using the DRASTIC method can be summarized into four phases: (i) data acquisition; (ii) the preparation of the map database; (iii) calculating the vulnerability index and; (iv) the classification index according to DRASTIC classes to determine vulnerability map.

- Definition of the DRASTIC data

The seven DRASTIC parameters are defined as follows:

- The Depth to water table: represents the vertical distance traversed by a contaminant over ground to reach the water table (saturated zone of the aquifer). The larger this depth is, the higher the contaminant puts enough time to reach the water table. It corresponds to the static level in drillings.

- Net recharge is the main vehicle for the transport of contaminant. The larger this recharge is, the more the risk of contamination is high. The net recharge determined in the zone considered is of 382 mm a year.

- The Aquifer media identifies the size of the saturated ground. It is involved in trapping the pollutant that can escape the power of absorption or adsorption of the soil. The larger the particle size is finer, trapping the pollutant is great. This parameter was identified from fifteen drill logs made in previous study.

- The Soil media controls the downward movements of contaminants. Indeed, the presence of fine materials (clays, silts and silt) and organic matter in the soil reduces the intrinsic permeability and retards the migration of contaminants, by physico-chemical processes (adsorption, ion exchange, oxidation, biodegradation). For this parameter, geotechnical map has been used to identify soil types in the study area.

- The Topography or slope (%) favors the runoff of surface waters at the expense of infiltration. The steeper the slope of the land, the more water runoff is important and therefore the groundwater contamination is low. The slope map is obtained by extraction of the contours of the DEM and transforming them into fashion TIN (Triangular Irregular Network) with the extension "3D Analyst" of GIS. The "Slope" of this extension allows converting the TIN in a slope map.

- The Impact of vadose zone is defined as the fraction between the water table and the surface where the pores are partially saturated with water. The permeability of the unsaturated zone controls the flow of pollutants and their arrival at the water table. Its impact is determined from the texture of the lands that constitute it. Percolation of the pollutant to the water table is so much greater that the texture is favorable. This parameter has been identified from the drill logs produced in the area.

- The hydraulic conductivity is the rate at which water flows through an aquifer. It refers to the ability of the geological formations of the aquifer to transmit water. More this setting, the greater the transfer of the pollutant is fast. The conductivity is determined by the ratio of the transmissivity and the estimated thickness of the water tables.

• Developing of the database maps

The map database is performed in the coordinate system of Universal Transverse Mercator (UTM), World Geodesic System (WGS) 84, Zone 30 and Northern Hemisphere. The geological map was georeferenced and used to scan the survey area using the functions "Georeferencing and Editor" of ArcInfo/ArcGIS 9.3. The method requires spatial interpolation or the DRASTIC parameters. Interpolation is the procedure for estimating the value of a variable in a particular place, from a number of measurement points to calculate each point in the study area [14] . In this study, the chosen interpolation method is the Inverse Distance Weighting (IDW) which is an estimate by linear combination of the values of the known points and is available in "Spatial Analyst" of ArcInfo/ArcGis 9.3".

• Distribution of water points

The study was conducted on 217 drillings which spatial distribution is shown in Figure 2. The values of DRASTIC parameters of each water point made it possible to carry out the interpolations by the chosen method for determining the thematic maps.

• Developing of the thematic maps

The database developed in the framework of this study has two components. A descriptive or alphanumeric component developed under Excel. These data were then imported into ArcInfo/ArcGIS through its ArcCatalog module intended to store and manage this numerical data. The raw values of each DRASTIC parameter were transferred in ArcMap where the interpolations have been performed by setting the spatial resolution (pixel) at 250 m × 250 m.

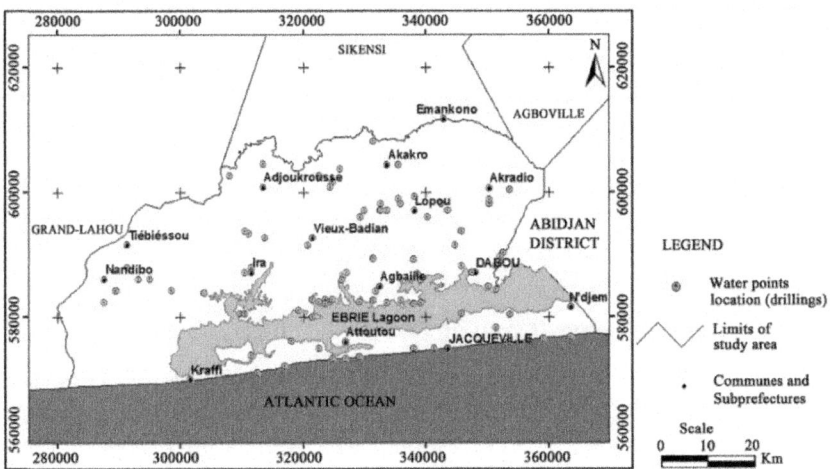

Figure 2: Water points distribution map in the study area.

Thereafter, the classifications according to the method were performed (Table 1) with the function "Reclassify of Spatial Analyst" of the GIS.

Each class is assigned an index named as a symbol (c), typically from 1 to 10 depending on the degree of its impact on the pollution of the aquifer. A weighting factor (p) or fixed multiplication factor of 1 to 5 is assigned to each parameter according to the method in order to relativize their respective importance in terms of vulnerability. This factor reflects the influence degree of each of them in the vulnerability assessment and is imposed by the method. DRASTIC parameter maps are obtained after this manipulations series in the GIS.

Thematic maps of various parameters are established based on DRASTIC ratings. The areas characterized by low scores indicate the favorable hydrogeological conditions to natural groundwater protection. However, for high ratings, these conditions are more and more critical as the vulnerability is high and groundwater is likely to be reached by possible pollution coming from the soil surface.

• Calculation of the DRASTIC vulnerability index

After mapping the parameters according to the rating system of the DRASTIC method, they were weighted. The weighting was to multiply each thematic map (DRASTIC ratings map) by its respective weight. The superposition of different layers (thematic maps) on a result-layer will allow calculating DRASTIC index map. The superimposed layers must obviously have the same cartographic characteristics namely the same projection system, the same units of length, the same geographical extent and also the same resolution as all calculations will be made on the same matrix [15] . The final vulnerability index or noted DRASTIC Vulnerability Index is equal to the weighted sum of seven parameters. The DRASTIC Vulnerability Index (DVI) is calculated by the Equation (1) [11] :

$$DVI = Dc \times Dp + Rc \times Rp + Ac \times Ap + Sc \times Sp + Tc \times Tp + Ic \times Ip + Cc \times Cp \quad (1)$$

With: D, R, A, S, T, I, C are the parameters mentioned above;

p is weighting factor given to each parameter;

c is rating assigned to each parameter.

- Classification index based on DRASTIC classes

The possible range of values vulnerability index is found between 23 and 226 in the case of the standard version. The values obtained are grouped in eight classes which each one corresponds to a vulnerability degree [11] . The reclassification of the index map within the ranges defined by the method used to develop the groundwater vulnerability to pollution map.

Table 1: Standardization of the parameters using the DRASTIC method [11] (modified)

DEPTH TO WATER TABLE (meter)			TOPOGRAPHY (slope %)		
Range (meter)	Rating	Weight	Range (%)	Rating	Weight
0 - 1.5	10		0 - 2	10	
1.5 - 4.5	9		3 - 6	9	
4.5 - 9	7		7 - 12	5	1
9 - 15	5	5	13 - 18	3	
15 - 23	3		>18	1	
23 - 30	2		IMPACT OF VADOSE ZONE		
>30	1		Range	Rating	Weight
NET RECHARGE (millimeter·year⁻¹)			Sandy clays	3	
Range (millimeter·year⁻¹)	Rating	Weight	Metamorphic rock Sandstone, Clay	4	
0 - 50	1		Armours and Gravel	5	
50 - 100	3	4	Ferruginous sandstone and coarse sands bitumen	6	5
100 - 180	6		Clayey Sand	7	
AQUIFER MEDIA			Sand and Gravel	8	
Range	Rating	Weight	coarse sands	9	
Sand and Gravel	8	3	HYDRAULIC CONDUCTIVITY (meter·day⁻¹)		
Clayey Sands	4		Range (meter·day⁻¹)	Rating	Weight
SOIL MEDIA			<4	1	
			4 - 12	2	
Range	Rating	Weight	12 - 29	4	3
Sands (10% to 45% of Clay)	7		29 - 41	6	
Organic matter (peats)	8	2	41 - 82	8	
Sands (0% to 8% of Clay)	9		>82	10	

Validation Methods of the Vulnerability Map: Sensitivity Analysis Tests

A sensitivity analysis by two tests was conducted to appreciate the effect of each parameter on the intrinsic sensitivity maps. These tests have been applied in recent studies [16] - [21] to analyze the reliability of vulnerability criteria and validate developed vulnerability maps. They are:

- Map removal sensitivity analysis: This test was developed by [22] . It identifies the sensitivity of the DRASTIC vulnerability map for each of the seven parameters. It is therefore to verify the relevance of the parameters in assessing the intrinsic vulnerability and to deduce the most representative in the application of the method. Thus, the degree of influence of a given parameter on the vulnerability index was evaluated by removal the parameter in overlay process. The comparison of the new vulnerability index with the original index provides a direct measure of the parameter influence. This measure corresponding to the sensitivity index (S) of the parameter is determined from the Equation (2):

$$S = \left| \frac{V}{N} - \frac{V_{xi}}{n} \right|$$
(2)

With:

S is the sensitivity index of the parameter;

V is the intrinsic vulnerability index of the method;

N is the total number of parameters used to calculate V;

V_{xi} represents the intrinsic vulnerability index obtained after removal of the parameter X.

- The single parameter sensitivity analysis: This test was developed by [23] . This test consists to assess the impact of each parameter in calculating the vulnerability index by comparing the calculated weight of the input parameter in each polygon with the theoretical weight assigned by the analytical model [2] . This is an analysis of the real influence of weighting compared to the weights assigned to each parameter. The real or effective weight (W) of a parameter is calculated from Equation (3):

$$W = \frac{Xr + Xw}{V} \times 100$$
(3)

With:

W is the effective weight of the parameter;

Xr and Xw are respectively the rate and the weight assigned to the parameter X and V intrinsic vulnerability index according to the method.

RESULTS AND DISCUSSION

- DRASTIC vulnerability index map

The superposition of the seven punctuated and weighted maps provides DRASTIC index map, which is carried out on the basis of a linear combination between different data values. Each pixel is characterized by a DRASTIC index value. The values of DRASTIC index are defined according to eight classes. These values represent the measurement of the hydrogeological aquifer vulnerability. The vulnerability increases with the index. These values range from 95 to 187 over the entire study area. This value range lies between the extreme values determined by the method that are 23 and 226. To the North and West of the study area, vulnerability index are relatively low and vary from 95 to 131. In the South and East of this zone, the indexes are increasingly important and vary from 131 to 187.

- Groundwater vulnerability map according to DRASTIC method

The application of the DRASTIC method combined to a GIS allowed obtaining the groundwater vulnerability to pollution map of the study area (Figure 3). This map is used to view the main risk areas linked to high index. This map is characterized by four vulnerability classes defined as follows:

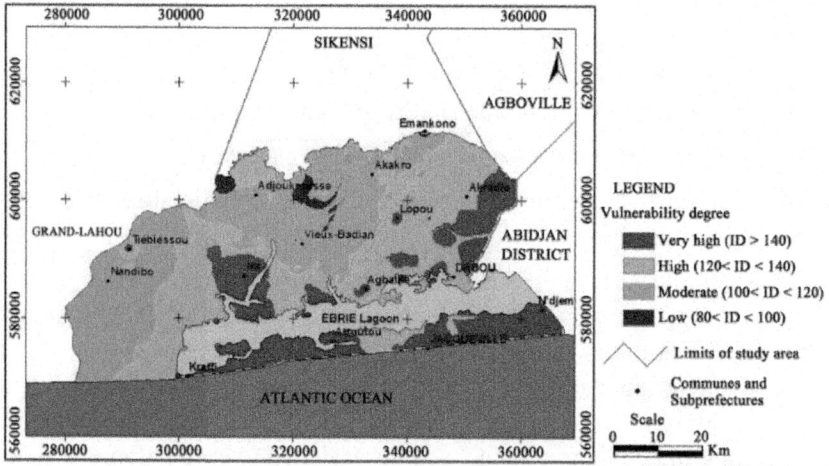

Figure 3: Groundwater vulnerability to pollution map of the study area.

- D > 140: very high vulnerability;
- 20 < ID < 140: high vulnerability;
- 00 < ID < 120: medium or moderate vulnerability;
- 0 < ID < 100: low vulnerability.

The analysis of this map shows a predominance of high and very high vulnerability classes where the index vary from 120 to more than 140 compared to moderate and low vulnerability classes where these index vary from 80 to 120. This map is in raster fashion characterized by a spatial resolution or pixel of 250 m × 250 m allowing knowing the area bounded by each class depending on the number of pixels given by the GIS.

These results show that: very high vulnerability class occupies 26.22% of the study area. This class is repre- sented in the South and East of the region where pollution risks incurred by the groundwater is very significant. The entire area contains almost all the department of Jacqueville where Quaternary formations are only represented. This extreme vulnerability can be explained by the depth to water table relatively very weak in the majority of drillings and where the clay fraction likely to attenuate pollution in the sandy formations is weak (0% - 8%). The high vulnerability class represents the modal class of the spatial distribution of vulnerability. It occupies 37.71% of the study area. This class occupies the central part of the study area. The vulnerability is moving towards the North-East and the South-West. The clay content likely to slow down the migration of the pollutant is slightly higher (10% to 45%) in sandy formations as in the previous case. The moderate vulnerability class is represented generally in North and West of the study area. It occupies 34.73% of the study area. The depth to water table becomes increasingly important with the presence of significant clay content in the formations. Hydraulic conductivity becomes low contrary to the first two classes. All these characteristics confer on the water tables a less significant vulnerability. Low vulnerability class is weakly represented in islands in North with only 1.34% of the study area. The area occupied by this class is almost negligible compared to those of other classes.

- Results of sensitivity analysis tests of DRASTIC vulnerability map

The maps resulting from the sensitivity analysis test by map removal for each of the seven parameters DRASTIC method are presented in Figures 4-10. These maps show a variation of the spatial distribution of the different classes and index calculated based on the removal of these parameters one. Table 2 summarizes the minimum, average and maximum partial index obtained.

The results in Table 2 indicate that the removal of the topography parameter or the hydraulic conductivity parameter of the overlay process were used to obtain more important average partial index values (respectively 121.48 and 123.53) compared to removal of each other parameters. However, removal the impact of vadose zone parameter of this process has to have the lowest average partial index value (97.50). The partial indexes determined were used to calculate subsequently the sensitivity index for each of the seven DRASTIC parameters (Table 3).

The analysis of the results in Table 3 shows that the depth to water table and the impact of the vadose zone have a strong influence on the vulnerability map with respective sensitivity index of 2.57 and 2.42. The topography and the hydraulic conductivity seem to have a moderate influence with respective sensitivity index of 1.58 and 1.26. The net Recharge, the aquifer media and the soil media with respective sensitivity index of 0.84, 0.52 and 0.59 to a lesser extent affect the DRASTIC vulnerability map index.

Table 2: Vulnerability partial index calculated by removing each of the seven DRASTIC parameters

Parameter Removal	Index			
	Minimum	Average	Maximum	Standard Deviation (SD)
D	66	110.36	144	10.87
R	71	106.98	163	14.29
A	71	108.88	163	16.95
S	81	115.53	173	10.07
T	87	121.48	177	14.22
I	60	97.50	147	10.70
C	89	123.53	167	13.37

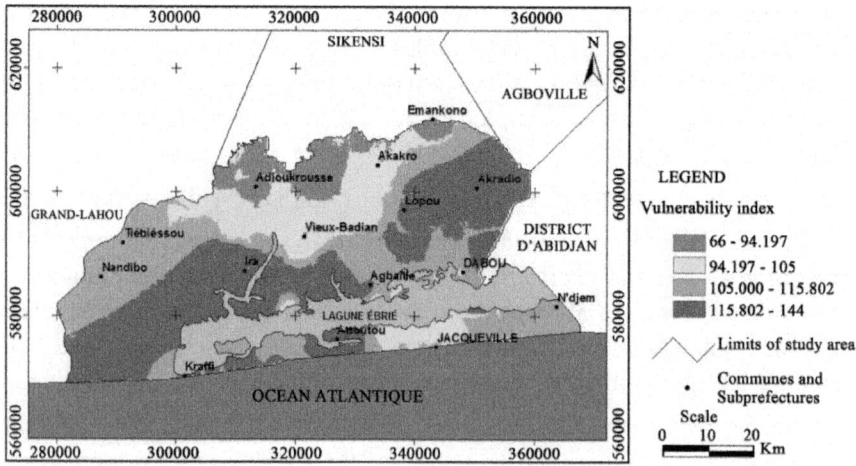

Figure 4: Sensitivity map by removing depth to water.

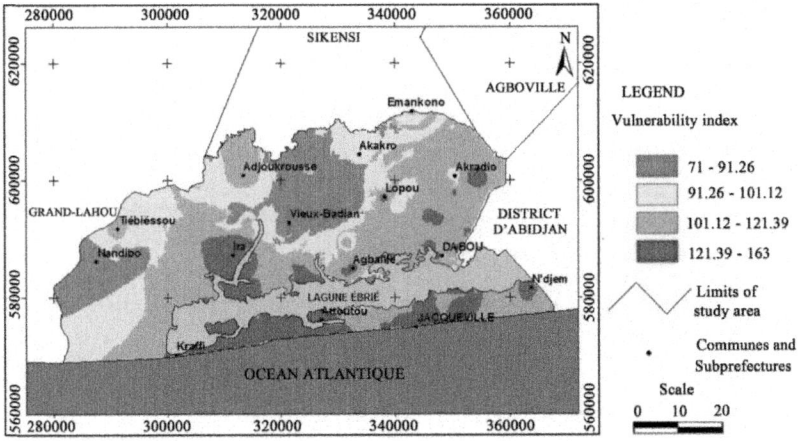

Figure 5: Sensitivity map by removing net recharge.

Table 3: Sensitivity index according to the map removal sensitivity analysis test for DRASTIC vulnerability map

Parameters	Sensitivity Index			
	S Minimum	S Average	S Maximum	Standard Deviation (SD)
D	0.28	2.57	2.71	0.26
R	0.45	0.84	1.74	0.30
A	0.45	0.52	1.74	0.75
S	0.07	0.59	2.12	0.27
T	0.93	1.58	2.79	0.30
I	2.21	2.42	3.57	0.29
C	1.12	1.26	1.92	0.15

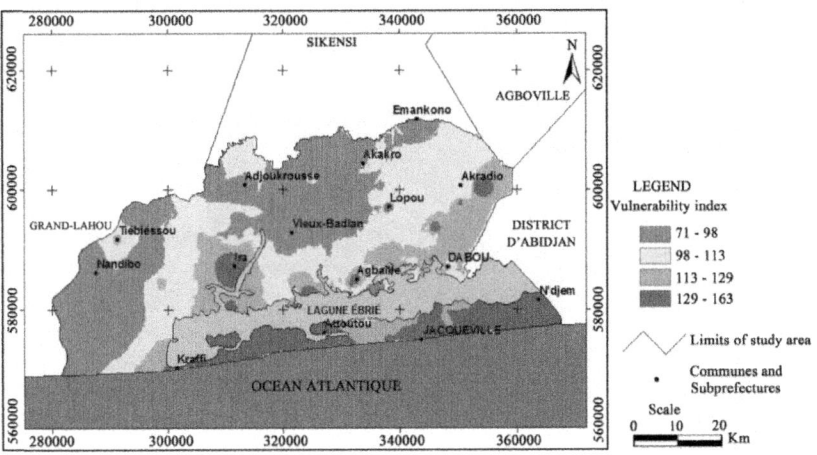

Figure 6: Sensitivity map by removing aquifer media.

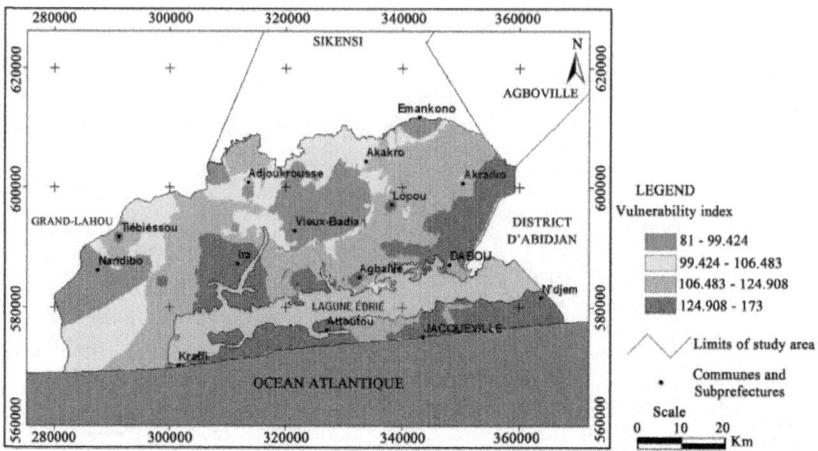

Figure 7: Sensitivity map by removing soil media.

The aquifer media is the parameter that has recorded the highest value of the standard deviation which indicates an important dispersion or a large variation of the results at the regional scale. Table 4 summarizes the results of the single parameter sensitivity analysis test.

The results in Table 4 show that the parameters such as the aquifer media (13.77%), the soil media (12.24%) and the impact of the vadose zone (22.96%) have their most important effective weight that their theoretical weight (13.04%, 8.7% and 21.74% respectively). For other parameters such as the depth of water table (20.20%), the net recharge (15.30%), the topography (4.28%) and the hydraulic conductivity (11.84%), the calculated effective weight are lower than the theoretical weight (21.74%, 17.39%, 4.35% and 13.04% respectively).

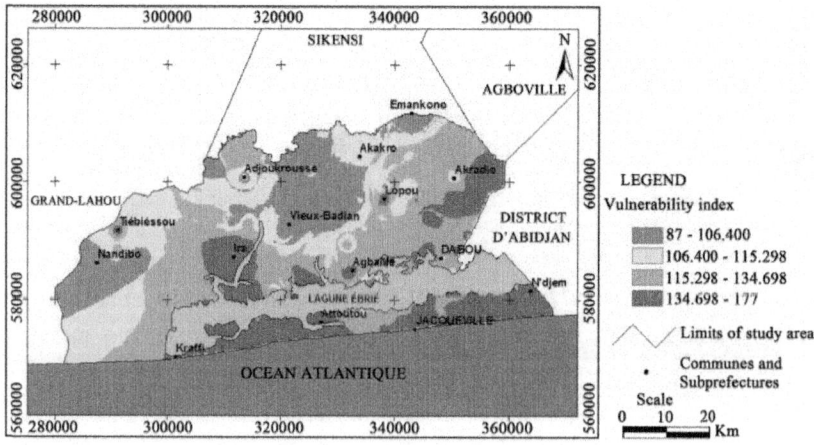

Figure 8: Sensitivity map by removing topography.

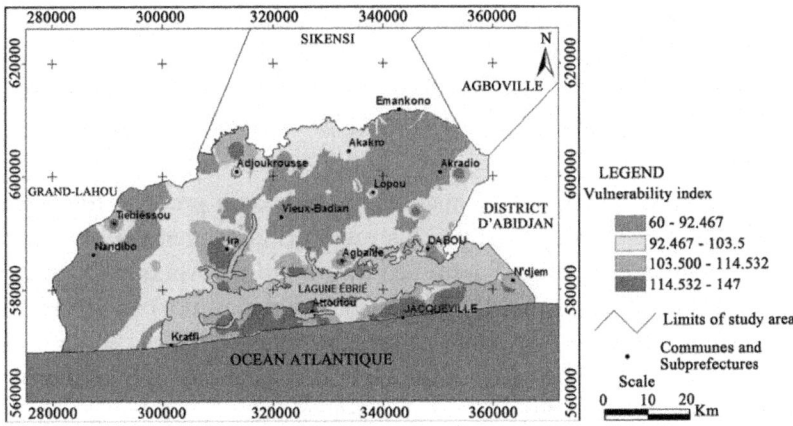

Figure 9: Sensitivity map by removing Impact of vadose zone.

Table 4: Effective weight compared to theoretical weight according to the single parameter sensitivity analysis test for the DRASTIC vulnerability map

Parameters	Weight				
	Theoretical Weight	Theoretical Weight (%)	Effective Weight (%)		
			W_{min}	W_{moy}	W_{max}
D	5	21.74	5.26	20.20	26.74
R	4	17.39	4.21	15.30	12.83
A	3	13.04	12.63	13.77	12.83
S	2	8.7	14.74	12.24	9.63
T	1	4.5	1.05	4.28	5.35
I	5	21.74	15.79	22.96	24.06
C	3	13.04	3.16	11.84	16.04

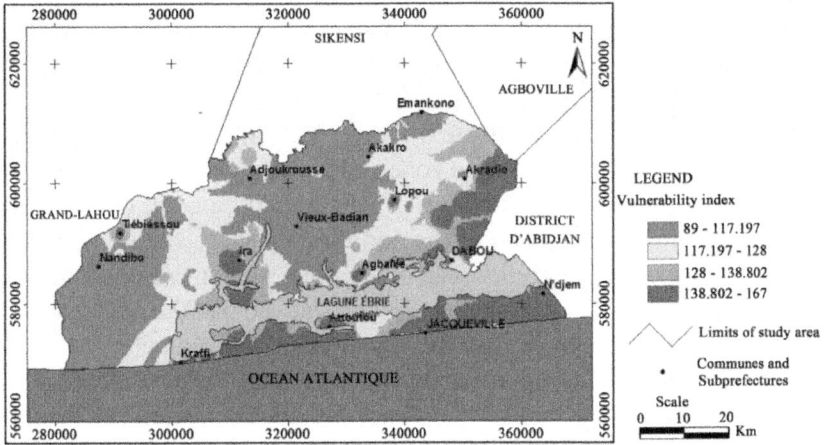

Figure 10: Sensitivity map by removing hydraulic conductivity.

Indeed, these results demonstrate that modeling in the GIS effectively reduces the subjectivity associated with the allocation of the weights to the various parameters using the DRASTIC method. These weights in fact are readjusted during modeling in the GIS to enable better assessment of the vulnerability.

The DRASTIC vulnerability map of the study area presents four classes: very high, high, moderate and low. This region is characterized by relatively low index compared to those determined by [24] in the Abidjan District (113 - 192) and those obtained by [25] in the Bonoua region (101 to 209) in Côte d'Ivoire. There exists all the same a bringing together of the index values. Indeed, it is in neighboring areas of the Ivorian coastal sedimentary basin that almost have the same hydrogeological characteristics justifying the bringing together of the index values in the assessment of the intrinsic vulnerability. The vulnerability index increases to the East or decreases to the West in this basin. These results show that the ground waters of the study area are less vulnerable than those of the Abidjan District and of the Bonoua region where calculated indexes are higher.

Moreover, it is clear from sensitivity analysis that the depth of water table and impact of the vadose zone have a high sensitivity on the maps of groundwater vulnerability in the study area. When each of these two parameters is removed from the overlay process, this leads to a significant decrease in vulnerability index. These results justify the allocation of greater weight to these two parameters in the application of the DRASTIC method and demonstrate the suitability of coupling GIS and DRASTIC method. These results are similar to those obtained by [20] who have shown that in addition to these parameters, the net recharge and the topography have a strong influence on vulnerability index in the Sub-Prefecture of Songon in the Abidjan District. The studies of [19] revealed that the topography and the depth of the water table are the parameters that have a strong sensitivity to the DRASTIC vulnerability index in the Aboisso District (Côte d'Ivoire) located on the rocky plinth. For [18] , it is the impact of the vadose zone and the hydraulic conductivity which significantly influences the groundwater vulnerability map in the Lifen Basin (China). For [16] , the aquifer media and the impact of vadose zone have a high sensitivity on vulnerability map of Russeifa area in Jordan. These same results differ from those obtained by [2] who showed that the net recharge, the soil media and the topography are the most influential parameters in assessing the aquifer vulnerability in Kakamigahara (central Japan), and that other parameters have a low to moderate impact. Indeed, the parameters that have a strong sensitivity to the DRASTIC vulnerability maps vary from one study to another. That is why [2] also point out that the variability of these parameters depends on the hydrogeological characteristics of each study area.

That same report is also valid in the comparison the theoretical and actual (or effective) weights to deduce the parameters actually involved and that have a significant impact in the assessment of vulnerability from the point of view of weighting. In this study, the parameters such as the soil media, the aquifer media and especially the impact of the vadose zone have been identified as having a significant impact in the vulnerability mapping. These results are in conformity with those of [26] and [20] who have shown that these parameters are, from this point of view, the most deter-mining in the assessing the groundwater vulnerability. [19] also showed that the net recharge made from these parameters while [18] is limited to parameters such as the aquifer media and the soil media. These results come nearer those of [27] and [16] that retained the parameters such as the depth of the water table, the aquifer media or the impact of the vadose zone respectively in the Turbio River valley (Mexico) and in the Russeifa region (Jordan). However, they differ from those of [2] which showed the significant impact of net recharge and hydraulic conductivity in assessing the aquifer vulnerability in Kakamigahara. In this study, the results of the both sensitivity analysis tests show that the impact of the vadose zone is the most important parameter in assessing the vulnerability in the study area. The groundwater vulnerability to pollution is linked to the nature of the sediments (hydrogeological formations) which condition the other parameters essentially the aquifer media and the soil media. According to [21] , the impact of the vadose zone reduces the subjectivity in assigning ratings or weight and increases the reliability in assessing of the groundwater vulnerability to pollution. Although, this parameter remains the most important, it requires all other parameters for a more objective assessment of vulnerability. According to [19] the removal of each of the seven parameters involves a sensitivity concerned with this parameter on the vulnerability map. That is why [23] confirmed that the seven DRASTIC parameters have to be considered very important in the assessment of the vulnerability. The DRASTIC vulnerability maps have already been the subject of several criticisms. For [23] these criticisms concern the inevitable subjectivity associated with the selection of seven parameters and assigning ratings and weights used to calculate the vulnerability index. Despite these limitations, the DRASTIC vulnerability map of the study area reflects in our opinion the ground realities. It can therefore be used as tool of decision-making aid for the management or the prevention of groundwater pollution risks. It helps to have an idea on sensitive areas which it will have to take into account during the regional planning [28] .

CONCLUSION

This study presents the assessment of the groundwater intrinsic vulnerability in the Dabou region from the DRASTIC method which requires seven hydrogeological parameters in its application. The established DRASTIC vulnerability map presents four classes: very high (26.22%) in the South and the East, high (37.71%) in the Center, the North-East and North-West, moderate (34.73%) in the North and West and low (1.34%) in the North. The indexes calculated during the establishment of this map vary from 95 to 187 indicating a greater vulnerability. The sensitivity analysis by map removal has shown that the impact of the vadose zone and the depth to water table has a much larger influence on the assessment of the groundwater intrinsic vulnerability in the study area. The vulnerability of groundwater in this region is strongly linked to these two parameters that allow for higher index therefore know the most sensitive areas. The single parameter sensitivity analysis shows that the aquifer media and the soil media have a more significant impact that the net recharge, the topography and the hydraulic conductivity in terms of weighting. The GIS used in this study by allowing the application of the DRASTIC method provide certain objectivity in the assessment of vulnerability. It allows easily per- forming the sensitivity analysis of groundwater vulnerability map of the Dabou region. This map is now a powerful tool of assistance in the prevention of the groundwater pollution risks. The vulnerability map constitutes a tool for monitoring human activities in the regional planning for the sustainable conservation the groundwater quality.

ACKNOWLEDGEMENTS

I would like to thank my supervisor Professor Jean Patrice JOURDA of Félix Houphouët-Boigny University (Côte d'Ivoire) for his support and guidance in completing this project. I would also like to thank Professor Jean Kan KOUAMÉ and Dr. Kouakou Serge DEH for their important contribution and all the members of the Laboratory of Remote Sensing and Spatial Analysis Applied to Hydrogeology for the success of this project.

REFERENCES

1.   Morris, B.L., Lawrence, A.R., Chilton, P.J.C., Adams, B., Calow, R.C. and Klinck, B.A. (2003) Groundwater and Its Susceptibility to Degradation: A Global Assessment of the Problem and Options for Management. Early Warning and Assessment Report Series, RS. 03-3, United Nations Environmental Program, Nairobi, Kenya, 126. http://www.unep.org/DEWA/water/groundwater/pdfs/Groundwater_INC_cover.pdf

2.   Babiker, I.S., Mohamed, M.A.A., Hiyanna, T. and Kato, A. (2005) A GIS Based DRASTIC Model for Assessing Aquifer Vulnerability in Kakamigahara Heights, Gifu Prefecture, Central Japan. Science of the Total Environment, 345, 127-140. http://dx.doi.org/10.1016/j.scitotenv.2004.11.005

3.   Dridi, L. and Schäfer, G.L. (2006) Vapor Flux of the Chlorinated Solvents Quantization from a Source Porous Aquifer to the Atmosphere: Bias Relative to the Non-Uniformity of the Water Content and the Non-Stationary of the Transfer. Mechanical Report, 334, 611-620.

4.   Melloul, A. and Collin, M. (1994) Water Quality Factor Identification by the "Principal Components" Statistical Method. Water Science Technology, 34, 41-50.

5.   Foster, S.S.D. (1987) Fundamental Concepts in Aquifer Vulnerability, Pollution Risk and Protection Strategy. Hydrological Resources Processes and Informations, 38, 69-86.

6.   Tapsoba, S.A. (1995) Contribution to the Geological and Hydrogeological Study of Dabou Region (Southern of Côte d'Ivoire): Hydrochemistry, Isotopic and Cationic Aging Index of Groundwater. PhD Thesis, 3rd Cycle National University of Côte d'Ivoire, 200.

7.   Statistic National Institute (SNI) (2010) General Census of Population and Housing. http://fr.wikipedia.org/wiki/Dabou http://fr.wikipedia.org/wiki/Jacqueville_.

8.   Tastet, J.P. (1979) Quaternary Sedimentary and Structural Environments of the Coastline of the Guinea Gulf (Côte d'Ivoire, Togo, Benin). PhD Thesis, (State) of Science University of Bordeaux 1, 181.

9.   Jourda, J.P. (1987) Contribution to the Geological and Hydrogeological Study of the Area of Greater Abidjan (Côte d'Ivoire). PhD Thesis, 3rd Cycle Scientific, Technical and Medical University of Grenoble, France, 319.

10.   Aghui, N. and Biémi, J. (1984) Geology and Hydrogeology of Groundwater in the Region of Abidjan and Contamination Risks. Annals of the National University of Côte d'Ivoire Series C, 20, 331-347.

11.   Aller, L., Bennett, T., Lehr, J., Petty, R.J. and Hackett, G. (1987) DRASTIC: A Standardized Method for Evaluating Potential Ground Water Pollution Using Hydrogeological Settings. EPA/600/2-87-035, 622 p.

12.   Schnebelen, N., Platel, J.P., Lenindre, Y. and Baudry, D. (2002) Groundwater Management in Aquitaine Year 5. Segment Operation,

Protection of the Water Table of the Oligocene in the Bordeaux Region. Report BRGM/RP-51178-FR, 20 p.

13. Mohamed, R.M. (2001) Evaluation and Mapping of Alluvial Aquifer Vulnerability to Pollution of the Plain of El Madher, Algerian North-East, Using the DRASTIC Method. Science and Planetary Change/Drought, 12, 95-101.

14. European Network for Research on Global Change (ENRICH) (2001) Climate-West Africa; A Network for: Harmonization of Climate Prediction for Mitigation of Global Change Impacts in Sudano-Sahelian. Energy, Environment and Sustainable Development Program Key Action 2. Global Change Climate and Biodiversity, University of East Anglia, Norwich.

15. Smida, H., Abdellaoui, C., Zairi, M. and Dhia, B.H. (2010) Mapping of Agricultural Areas Vulnerable to Pollution by the DRASTIC Method Coupled with a Geographic Information System (GIS): Case the Chaffar Water Table (South of Sfax, Tunisia). Science and Planetary Change/ Drought, 21, 131-146.

16. El-Naqa, A., Hammouri, N. and Kuisi, M. (2006) GIS-Based Evaluation of Groundwater Vulnerability in the Russeifa Area, Jordan. Revisita Mexicana de Ciencias Geológicas, 23, 277-287.

17. Akbari, G.H. and Rahimi-Shahrbabaki, M. (2011) Sensitivity Analysis of Water at Higher Risk Subjected to Soil Contamination. Computer Methods in Civil Engineering, 2, 83-94.

18. Samake, M.H., Tang, Z., Hlaing, W., Ndoh, M.I., Kasereka, K. and Balogun, W. (2011) Groundwater Vulnerability Assessment in Shallow Aquifer in Linfen Basin, Shanxi Province, China Using DRASTIC Model. Journal of Sustainable Development, 4, 53-71. http://dx.doi.org/10.5539/jsd.v4n1p53

19. Doumouya, I., Dibi, B., Kouame, I.K., Saley, B., Jourda, J.P., Savane, I. and Biémi, J. (2012) Modeling of Favorable Zones for the Establishment of Water Points by Geographical Information System (GIS) and Multicriteria Analysis (MCA) in the Aboisso Area (South-East of Côte d'Ivoire). Environmental Earth Science, 67, 1763-1780. http://dx.doi. org/10.1007/s12665-012-1622-2

20. Dibi, B., Kouame, K.I., Konan-Waidhet, A.B., Savane, I., Biémi, J., Nedeff, V. and Lazar, G. (2012) Impact of Agriculture on the Quality of Groundwater Resources in Peri-Urban Zone of Songon (Côte d'Ivoire). Environmental Engineering and Management Journal, 11, 2173-2182. http://www.academia.edu/18674209/Impact_of_agriculture_on_the_

quality_of_groundwater_resources_in_peri-urban_zone_of_Songon_
Cote_D_ivoire_

21. Kouamé, K.I., Douagui, G.A., Koffi, K., Dibi, B. and Kouassi, K.K. (2013) Modeling of Quaternary Groundwater Pollution Risk by GIS and Multicriteria Analysis in the Southern District Share of Abidjan (Ivory Coast). Journal of Environmental Protection, 4, 1213-1223. http://dx.doi.org/10.4236/jep.2013.411139

22. Lodwick, W.A., Monson, W. and Svoboda, L. (1990) Attribute Error and Sensitivity Analysis of Map Operations in Geographical Information Systems: Suitability Analysis. International Journal of Geographical Information Systems, 4, 413-428. http://dx.doi.org/10.1080/02693799008941556

23. Napolitano, P. and Fabbri, A.G. (1996) Single Parameter Sensitivity Analysis for Aquifer Vulnerability Assessment Using DRASTIC and SINTACS. In: Kovar, K. and Nachtnebel, H.P., Eds., HydrolGis Application of Geographic Information Systems in Hydrology and Water Resources Management, IAHS Publication, Wallingford, 559-566.

24. Kouamé, K.J. (2007) Contribution to the Integrated Water Resources Management (IWRM) of the Abidjan District (South Côte d'Ivoire): Tools for Decision for the Prevention and Protection of Groundwater against Pollution. PhD Thesis, University of Cocody, Abidjan, 226 p.

25. Aké, G.E. (2009) Impacts of Climate Variability and Anthropogenic Pressures on Water Resources of Bonoua Region (South-East of Côte d'Ivoire). PhD Thesis, University of Cocody, Abidjan, 201 p.

26. Al-Hanbali, A. and Kondoh, A. (2008) Groundwater Vulnerability Assessment and Evaluation of Human Activity Impact (HAI) within the Dead Sea Groundwater Basin, Jordan. Hydrogeology Journal, 16, 499-510. http://dx.doi.org/10.1007/s10040-008-0280-7

27. Ramos, L.J.A. and Castillo, R.R. (2003) Aquifer Vulnerability Mapping in the Turbio River Valley, Mexico: A Validation Study. Geofísica Internacional, 42, 141-156.

28. Jourda, J.P. (2005) Application Methodology Remote Sensing Techniques and Geographical Information Systems to the Study of Fissured Aquifers of West Africa. Hydrotechnic Concept of Space: The Case of Côte d'Ivoire Testing Areas. PhD Thesis, University of Cocody, Abidjan, 430 p.

Chapter 6

A METHODOLOGY TO DEVELOP THE INTEGRATION OF THE ENVIRONMENTAL MANAGEMENT SYSTEM WITH OTHER STANDARDIZED MANAGEMENT SYSTEMS

Manuel Ferreira Rebelo[1], Gilberto Santos[1,2], Rui Silva[1]

[1]CLEGI, Lusíada University, Vila Nova de Famalicão, Portugal
[2]College of Technology, Polytechnic Institute of Cávado and Ave, Barcelos, Portugal

ABSTRACT

Traditionally the global management system of an organization is frequently split into a number of individual management systems that are defined and implemented according to specific man- agement systems standards (MSSs) as well as managed independently. The individual implemen- tation of MSSs is an option that leads to several inefficiencies and sub-optimization of the global management system of an organization. As referred by ISO [1] the interested parties' require- ments increase. A more effective and efficient option for an organization is to integrate, into an integrated management system (IMS), the implementation and management of requirements of multiple MSSs. Certain difficulties are associated to the structuring process, implementation, verification, evaluation, improvement and progressive development of an IMS in the organizations. Several scholars have proposed various theoretical approaches regarding the integration of individual management systems (MSs) leading to the conclusion that there is not a common practice for all organizations as they encompass different characteristics. This paper aims to present and justify a designed methodology to be used by organizations to support the integration of various MSs. Among them are highlighted: the Environmental Management System (EMS) according ISO 14001 [2] , the Quality Management System (QMS) according ISO 9001 [3] , and the Occupational Health and Safety Management System (OH & SMS) according OHSAS 18001 [4] . The methodology was designed in the context of a Portuguese company, on sequence of an organizational diagnosis and a research that was performed through a questionnaire. The strategy and the research methods took into consideration the case study.

INTRODUCTION

Due to the demand of the market itself or by other internal reasons, there are many organizations that implement different standardized MSs [5] . Standards arise through the development of detailed descriptions of particular characteristics of a product or service by experts from companies and scientific institutions [6] . According to ISO [7] , the domain of standardized management systems (MSs) has expanded greatly over the last years and nowadays there exist a relevant number of MSSs for individual MSs, which apply to any type of organization independently of its external and internal context. The objective of the development of standards is to support both individuals and companies when procuring products and services [6] . According to ISO publication [1] a common objective of MSSs is to assist organizations to manage the risks associated with providing products and services to customers and other interested parties.

As the number of MSSs versus standardized MSs increases, their integration becomes a necessity [1] [8] . In the literature and in MSSs are presented several definitions/descriptions of the concept of management system (MS) and integrated management system (IMS). Empirical studies were conducted with the aim of the integra- tion of individualized MSs around the world [9] and several tangible and intangible gains for organizations, as well as to their internal and external interested parties, are achieved with the integration of the individual stan- dardized MSs [10] - [15] , among others. Integration of MSs promotes synergies and cost savings, as well as a re- duction of the time spent when managing the systems [16] . Olaru [17] summarized forty benefits that an or- ganization can gain from the implementation of an IMS. Organizations therefore need a framework to integrate these MSs and facilitate their contribution to the operation of the overall business MS [18] . On the other hand an essential element in the strategy of any organization is the minimization of business risk to a level that ensures the security market [19] . According to Suditu [20] for organizations that want to survive and compete in the ac- tual market, it is necessary to continually improve performance of their business and MSs in a sustainable way, taking in consideration the necessities of the interested parties. In turn, according to Oliveira [21] , integration is justified as a function of the benefits that it provides; certifiable management systems that work separately are more bureaucratic and costly, and generate poorer results than those obtained employing integration. Study conducted by Simon [22] concluded that Organizations prefer integration of MSs to managing them separately and the integration of systems is one of the major strategies for ensuring survival and savings for the organizations of the sample.

Quality is no longer, as formerly, a redundant and restricted concept and must be managed in a global per- spective and of sustainability not only focused on satisfying customers, but on a whole range of interested par- ties [11] . Are example, those identified in the ISO 9004 [7] , several others exist. On the other hand, the increas- ing global competition potentiates, therefore, an increase in the expectations of all the interested parties of or- ganizations. On Table 1 there are listed several interested parties and associated needs and expectations to be satisfied by organizations according to the requirements of related MSSs.

So, more than ever, business sustainability gains increased importance and focus is shifting from their finan- cial results. These results will not verify if that focus does not prioritize also, the satisfaction in a balanced and integrated fashion of customers and others relevant interested parts, that are clearly and objectively the employees for example. In this context of real and new paradigms of management—the Global Quality Management—it is required a constant search for Business Excellence [11] .

Hence, in a not distant past, some organizations in Portugal and other countries, although in a small percent- age, began to integrate their individual standardized management systems like: EMS; QMS; OH & SMS; CSRMS- Corporate Social Responsibility Management System, among others. For this purpose, organizations began to conceive integrated procedures in order to make the integration of two systems (QMS & EMS or EMS & OHSMS) and whenever possible, the three standardized Management Systems: EMS, QMS, and OHSMS [23] . This reveals the growing interest that has been demonstrated by organizations in the adoption of the MSSs ISO 14001, ISO 9001 and OHSAS 18001.

Table 1: Examples of interested parts and their needs and expectations

Interested Parties	Needs and Expectations
Customers	Quality; price; delivery performance of products
Owners/shareholders	Sustained profitability; transparency
People in the organization	Good work environment; job security; recognition and reward
Suppliers and partners	Mutual benefits and continuity
Society	Environmental protection; ethical behaviour; compliance with statutory and regulatory requirements
Competitors	Ethical behaviour; fair competition; zero ethical faults
Government	Attractive employer; business continuity; compliance with statutory and regulatory requirements; energy efficiency; mutual benefits; on time payment of taxes and others fees; risk management; sustained profitability; transparency
Labor unions	
Regulators	

Note: Adapted and upgraded from ISO [7].

On the other hand, the integration of MSs, supported by those MSSs in a single system, taking into account the correspondence and the level of compatibility between them and potential tangible and intangible gains resulting from this integration will be an added value that organizations cannot ignore [11] . On the other hand, regulations based on ISO 9000 have been created to guide companies in developing systems for management and prevention of worker risks. Annex A and B of ISO 9001 [3] gives various clauses and subclauses related to the necessary elements of this standard [24] .

Human resources are the main subject of an organizations activity. Quality improvement and the efficiency of the organization activity depend greatly on the quality of human resources [25] . Top management support and commitment are thus essential for the initiation of the integration process, completed and subsequently main- tained. Managers consequently need to recognise that for the IMS to be implemented and maintained, they must continuously push it forward [26] . One interesting finding in research conducted by Alolayan [27] was the fairly strong correlation between top management commitment and the adoption of continuous improvement programs in the work organization. Sustainable management is the combination of management theory and the concept of sustainable development [28] , cited by Tsai and Chou [29] and according to Salomone [30] , a cultural shift is underway and the number of companies with more than one certification is constantly increasing. Many of them are advancing towards integration.

Within the past decade, the application of certification has spread from documenting quality standards to ad- ditional areas, including the management of occupational health and safety (OHS) [31] . To satisfy the require- ments of each standardized MSs like: EMS, QMS, and OHSMS organizations have to assure a lot of docu- mented procedures and other documentation, checking processes and associated records forms, among other several paperwork [32] . As stated by Zeng [33] major problems for enterprises to operate multiple parallel MSs include: it causes complexity of internal management, it lowers management efficiency, it incurs in cultural in- compatibility and it causes employee hostility and increases management costs. In a recent past, in Portugal and other countries, some companies have begun to integrate their individual MSs [34] . Figure 1 [9] presents the evolution on the number of certifications of integrated MSs (Quality, Environment, and OH & S) for 2007, 2010, and 2011 in Portugal.

IMSs and its certification are increasingly used by organizations namely to document and develop optimized conformance and business risk management, in a lean and sustainable way, in a variety of different management areas considering, in general, its internal and external context and the needs and expectations of interested par- ties, in particular. According to Jørgensen [37] the third and most ambitious level, the integration, concerns the creation of a culture of learning, focus on interested parties, continuous improvements, and synergies between the subject areas. This integration creates a sound basis for working towards a more sustainable MS.

APPROACH TO STRUCTURING A METHODOLOGY TO DEVELOP AN IMS

The standards ISO 14001 [2] , ISO 9001 [3] and OHSAS 18001 [4] , and the identification of common areas and requirements versus correspondences between them allowed to structure from the existing individual MSs in organizations, a methodology to develop the integration of the EMS with others standardized MSs like QMS and OH & SMS.

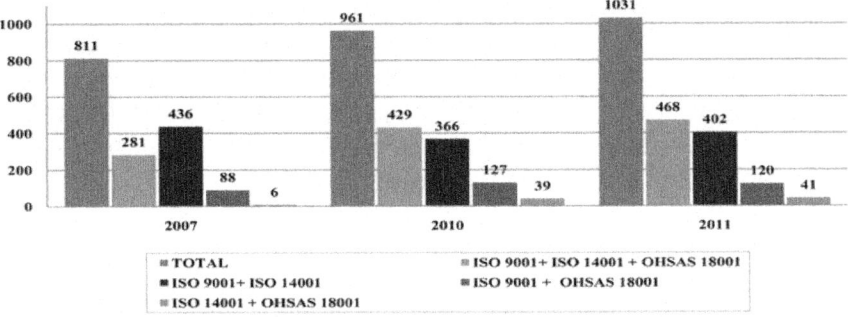

Figure 1: Evolution on the number of certified companies in Portugal related to integrated MSs (Quality, Environment, and OH & S) for 2007, 2010, and 2011. Source: Data from Portuguese Certified Companies Guide [35] [36].

There is a convergence in terms of the management model of the standards: ISO 14001 [2] , ISO 9001 [3] , OHSAS 18001 [4] , specifically in terms of the objectives associated with process efficiency, and to the fact that the three MSSs are supported on the continuous improvement principle of the Deming's Cycle-PDCA (Plan/Do/Check/Act), as described in Table 2.

The ISO 14001 [2] specifies requirements for an EMS to enable an organization to develop and implement a policy and objectives which take into account legal requirements and other requirements to which organizations subscribe, and information about significant environmental aspects. It applies to those environmental aspects that organizations identify as those which can be controlled and those which it can be influenced; the ISO 9001 [3] specifies requirements for a QMS where an organization: a) needs to demonstrate its ability to consistently provide products that meet customer and applicable statutory and regulatory requirements; and b) aims to en- hance customer satisfaction through the effective application of the system, including processes for continual improvement of the system and the assurance of conformity to customer and applicable statutory and regulatory requirements and the OHSAS 18,001 [4] specifies requirements for an OH & SMS to enable an organization to develop and implement a policy and objectives which take into account legal requirements and information about OH & S risks. So, there are several MSSs domains with potential for the effective integration of EMS with QMS and OH&SMS and integration gives a true usefulness and added value to the organization's business, more easier manageable and securely enhances the improvement of conditions in organizations in terms of management, the prevention component of EMS; QMS and OH & SMS [11] . According to Santos [23] , the com- patibility among different MSSs for an effective integration of the individual standardized MSs should be done in moderation and Almeida [38] [39] states that the success of the integration of the MSs is significantly related to the true motivations that leads organizations to integration. To achieve sustained success, top management should establish and maintain a mission, a vision and values for the organization. These should be clearly under- stood, accepted and supported by people in the organization and, appropriate to other interested parties [7].

MATERIALS AND METHOD

A preliminary investigation was conducted in the business environment, in a company, localized in the northern region of Portugal. Over the years the company—a SME, has been progressively adopting, in whole or in part, individualized MSSs and others specifications to implement independent MSs. Relevance to the ISO 14001 [2] for the EMS; ISO 9001 [3] for the QMS; OHSAS 18001 [4] for the OH & SMS, and ISO/International Electrotech- nical Commission (IEC) 17025 [40] for Laboratories MS and Accreditation.

While it was imperative to assess the perception of employees of the Company on the methodology, structur- ing, implementation and evaluation of the integration model of IMS and its validation in a real business envi- ronment, it was developed an internal investigation by questionnaire, previously tested and validated, to a repre- sentative sample, of employees. Figure 2 shows the distribution of collaborators surveyed from the different le- vels of the organizational structure.

Number (percentage) of Collaborators by Hierarchical Level

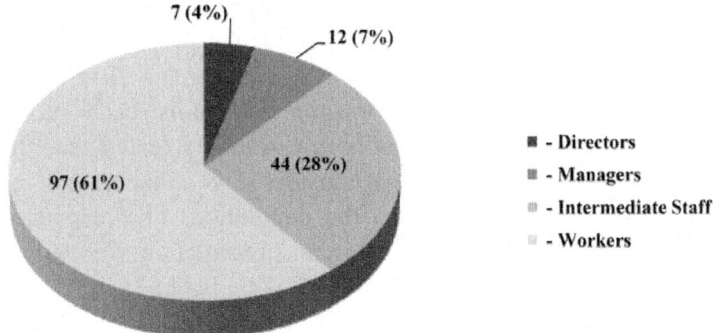

7 (4%)

12 (7%)

44 (28%)

97 (61%)

■ - Directors
■ - Managers
▩ - Intermediate Staff
▩ - Workers

Figure 2: Distribution of collaborators by hierarchical level. Source: [15].

Table 2: PDCA (Plan/Do/Check/Act) convergence at the level of the management model by the correspondent MSSs-ISO 14001 [2] , ISO 9001 [3] , and OHSAS 18001 [4]

PDCAI Cycle	Management Systems Standards		
	ISO 14001 [2]	ISO 9001 [3]	OHSAS 18001 [4]
PLAN To understanding the context of the organizations, to establish the necessary objectives and plans to achieve them, the processes, and their sequences and interconnections, as well as the required criteria and methods to guarantee their control and effectiveness	**PLANNING** Environmental targets and objectives; identification of environmental aspects and assessment of potential impacts; extent to which material and financial resources are affected; competence, training and awareness of the Human Resources; planning of the activities	**RESOURCE MANAGEMENT** Personnel competence, training and awareness; material resources; infrastructure and work environment	**PLANNING** OH & S objectives; hazard identification and risk assessment; extent to which material and financial resources are affected; training, awareness and competence of the Human Resources; communication, participation and consultation; planning of the activities
DO Implement the processes, ensuring the availability of support resources and appropriate documentation and the required documented information for realization and monitoring	**IMPLEMENTATION AND OPERATION** Environmental management programs; control of documents; operational control; emergency preparedness and response	**PRODUCT AND/OR SERVICE REALIZATION** Key processes; purchasing; service provision; measurement and monitoring equipment	**IMPLEMENTATION AND OPERATION** OH & S management programs; control of documents; operational control; emergency plans
CHECK Perform the measurements of the processes and monitoring their performance. Analyze and assess on a regular basis the obtained results	**CHECKS AND CORRECTIVE ACTIONS** Monitoring and measurement; compliance-nonconformities assessment, corrective and preventive actions; records control; internal audits	**MEASUREMENT, ANALYSIS AND IMPROVEMENT** Client and employee satisfaction; corrective actions for non-conformities; preventive actions; internal audits	**CHECKS AND CORRECTIVE ACTIONS** Monitoring and measurement; accidents, nonconformities and corrective and preventive actions; records control; internal audits
ACT Take the needed actions to meet the objectives, achieve the expected results and to encourage improvements	**POLICY** Environmental aspects; legal and other requirements; environmental policy	**MANAGEMENT RESPONSIBILITY** Quality policy Decentralized management; communication and information	**POLICY** Hazards and risks; legal and other requirements; OH & S policy
IMPROVE Continual improvement and innovation of MSs processes	Continual improvement and innovation of the MSs, with the main objectives of its optimization and satisfaction of all interested parties with focus on them for consequent development and sustained success of the organizations.		

Note: Adapted and upgraded from [11] [13].

Note: Adapted and upgraded from [11] [13] .

It was considered a Likert scale on the questionnaire with five levels: 1) irrelevant; 2) not so relevant; 3) rele- vant; 4) ery relevant; 5) determinant. The population was the total of the collaborators of the company from whom the objective was to draw conclusions about the project objectives and the issues that are being re- searched. After the questionnaire had been tested and improved in some of its questions, it was sent by e-mail to each one the company collaborators of the sample that had been carefully selected according to their position in the hierarchy. The sample considered 49 collaborators, representing 30.62% of the total collaborators—the population. The responses rate was 86%. In the data collection, analyses and presentation were considered the guidelines of the Portuguese standard—NP 4463 [41] .

There were considered four main questions: Question 1—importance of the twelve factors identified as moti- vation for the implementation of the IMS; Question 2—influence of nine identified interested parties on the performance and evolution of the IMS; Question 3—main difficulties in a group of seven potentials, in the con- text of the development and implementation of the IMS model. Question 4—potential benefits with the imple- mentation of the IMS-QES.

The main final objective of this preliminary research was to contribute to the integration of EMS, QMS, and OH&SMS in a specific organizational context, supported on a structured methodology of development of the integration, implementation and evaluation of the designed integration model and its validation. The survey re- sults, by them self, justify, validate and prioritize enormously the structure of the designed methodology and model of IMS.

MATRIX OF COMPATIBILITY OF REQUIREMENTS-SUP-PORT TO INTEGRATION

One of the activities that forms part of the scope and objectives of this preliminary research to which we have paid particular attention is the compatibility of the requirements of the MSSs, in context and framework of the characterization of the company's situation, backed up by an analysis of these MSSs. This compatibility, as pre- sented in Table 3, represent a starting point for consequents activities of integration, simplification and optimi- zation, to achieve a level of the strictly necessary and consequently the three sub MSs—EMS, QMS, and OH & SMS are integrated to the maximum extent possible.

On the matrix of Table 3 it is shown the requirements of the ISO 14001 [2] , ISO 9001 [3] , and OHSAS 18001 [4] , as well as the established correspondences, made them compatible with each other and associated with the phases of the cycle PDCA-"Plan-Do-Check-Act". This matrix orientate and align the organizational structure of the enterprise in the same direction and in addition creates a structured and useful referential methodology of work to support an effective alignment and correspondences of the sub management systems of Environment, Quality, and Safety with consequent compatibilities between each other, for consequent design and implementa- tion of the IMS. From this matrix it can also be depicted a correspondence with the PDCA cycle, in this circum- stance for the IMS, as well as a set of stages (1.1; 2.1...2.4; 3.1...3.7; 4.1...4.6 and 5.1) associated with each of the phases of the PDCA cycle.

MODEL OF DEVELOPMENT OF THE IMS

The continuous improvement of the global performance of an organization is an objective always present in the development of an IMS. The organization should therefore potentiate for each stage: Plan, Do, Check, Act, to be carefully and methodically analyzed in their differences that effectively can be observed in terms of MSSs re- quirements under clauses equivalent involved and for each phase and each stage of development of the IMS, according to the model of Figure 3 [11] [12] , to ensure its compliance and evidence of it, in full conformity.

Organizational diagnosis is an exercise done to check an organization's current health. A complete diagnosis not only checks the current health, but also suggests corrective measures [42] . So, first of all, the understanding of context of an organization, and definition, approval and communication of the integrated management policy, attentive that is common requirement to the different normative references. The leadership of top management is extremely important for the improvement of management quality. If the quality of top management is bad even if the management system of organization is good, it cannot continue to supply good outcomes such as products or services. As a result, it is thought that consequent profit is not ensured [25] . So, top management should demonstrate a strong commitment leadership and personal involvement through a defined Strategy, Policy, Ob- jectives and Targets for Quality, Environment and Safety, as well as, to make available all the needed resources to achieve the objectives [14] . The policy has to be coherent with the Mission, Vision, and Values of the com- pany, these supported on a strategy and specific objectives which in turn, support the implementation of that policy and its consequent effectiveness and continual improvement. The planning of activities in the aim of the Integrated Management System- Phase I—PLAN, is perhaps the most important of all [11] [12] . In fact, a neglected planning, including the non-identification of critical success factors, will lead to inefficiencies that can be translated into potential deviations to the objectives and consequent unsuccessful implementation of IMS.

Table 3: Matrix of compatibility of MSSs requirements and of support to the integration of the individual standardized MSs: EMS, QMS, and OH & SMS

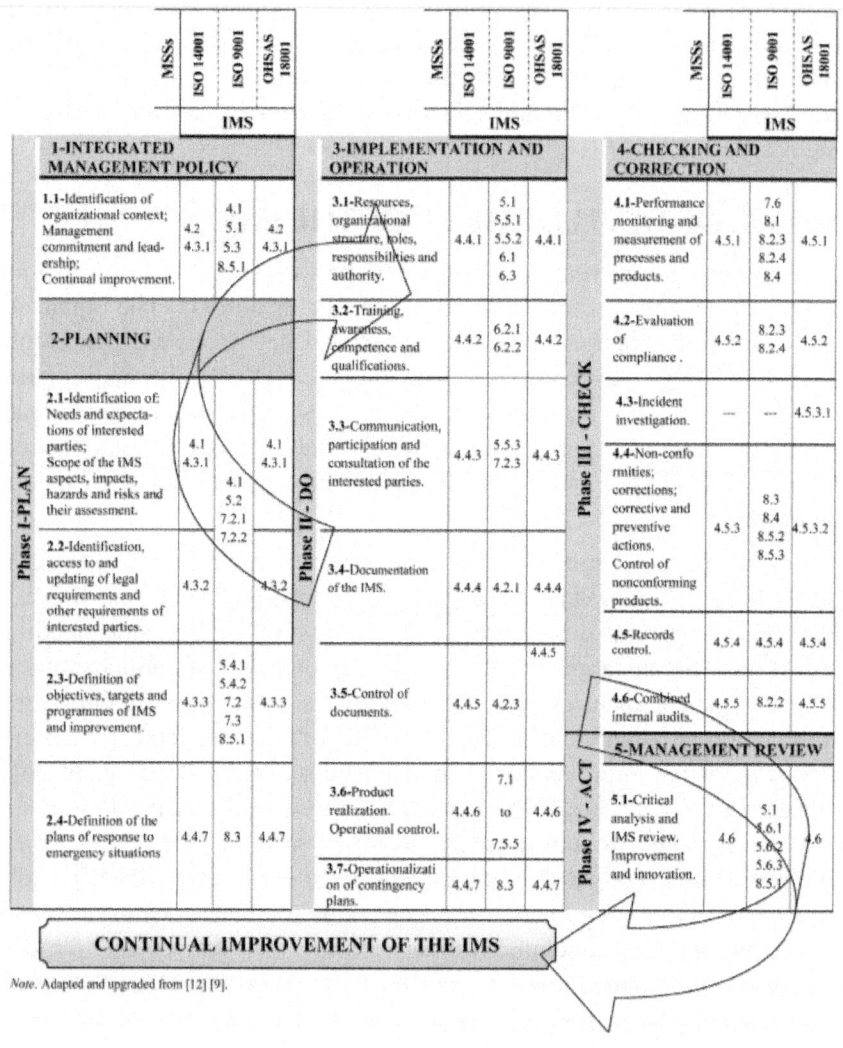

	MSSs	ISO 14001	ISO 9001	OHSAS 18001		MSSs	ISO 14001	ISO 9001	OHSAS 18001		MSSs	ISO 14001	ISO 9001	OHSAS 18001
			IMS					IMS					IMS	
1-INTEGRATED MANAGEMENT POLICY					**3-IMPLEMENTATION AND OPERATION**					**4-CHECKING AND CORRECTION**				
1.1-Identification of organizational context; Management commitment and leadership; Continual improvement.		4.2	4.1 5.1 5.3 8.5.1	4.2	3.1-Resources, organizational structure, roles, responsibilities and authority.		4.4.1	5.1 5.5.1 5.5.2 6.1 6.3	4.4.1	4.1-Performance monitoring and measurement of processes and products.		4.5.1	7.6 8.1 8.2.3 8.2.4 8.4	4.5.1
		4.3.1		4.3.1										
2-PLANNING					3.2-Training, awareness, competence and qualifications.		4.4.2	6.2.1 6.2.2	4.4.2	4.2-Evaluation of compliance .		4.5.2	8.2.3 8.2.4	4.5.2
2.1-Identification of: Needs and expectations of interested parties; Scope of the IMS aspects, impacts, hazards and risks and their assessment.		4.1 4.3.1	4.1 4.1 5.2 7.2.1 7.2.2	4.1 4.3.1	3.3-Communication, participation and consultation of the interested parties.		4.4.3	5.5.3 7.2.3	4.4.3	4.3-Incident investigation.		---	---	4.5.3.1
										4.4-Non-conformities; corrections; corrective and preventive actions. Control of nonconforming products.		4.5.3	8.3 8.4 8.5.2 8.5.3	4.5.3.2
2.2-Identification, access to and updating of legal requirements and other requirements of interested parties.		4.3.2		4.3.2	3.4-Documentation of the IMS.		4.4.4	4.2.1	4.4.4					
									4.4.5	4.5-Records control.		4.5.4	4.5.4	4.5.4
2.3-Definition of objectives, targets and programmes of IMS and improvement.		4.3.3	5.4.1 5.4.2 7.2 7.3 8.5.1	4.3.3	3.5-Control of documents.		4.4.5	4.2.3	4.4.5	4.6-Combined internal audits.		4.5.5	8.2.2	4.5.5
										5-MANAGEMENT REVIEW				
					3.6-Product realization. Operational control.		4.4.6	7.1 to 7.5.5	4.4.6	5.1-Critical analysis and IMS review. Improvement and innovation.		4.6	5.1 5.6.1 5.6.2 5.6.3 8.5.1	4.6
2.4-Definition of the plans of response to emergency situations		4.4.7	8.3	4.4.7	3.7-Operationalization of contingency plans.		4.4.7	8.3	4.4.7					

Phase I-PLAN Phase II-DO Phase III - CHECK Phase IV - ACT

CONTINUAL IMPROVEMENT OF THE IMS

Note. Adapted and upgraded from [12] [9].

It is therefore fundamental to invest resources and expertise at this stage, via a thorough and careful work, in order to respond effectively, within an integrated approach, to all interested parties requirements arising from the in- volved standards and others applicable specifications, including legal requirements, in this phase of the planning of the IMS with particular focus on environmental issues, customer satisfaction, and occupational health and safety of the collaborators and their families [11] [12] .

Following is the Implementation and Operation—"Do", the organization should, in this Phase II—DO, pro- mote the "Make/Do" in coherence with what was previously planned, attentive the scope of the IMS. Corresponds mainly to clauses: 4.4—Implementation and operation of ISO 14001 [2] , 7—Product Realization, of ISO 9001 [3] , and 4.4—Implementation and operation of OHSAS 18001 [4].

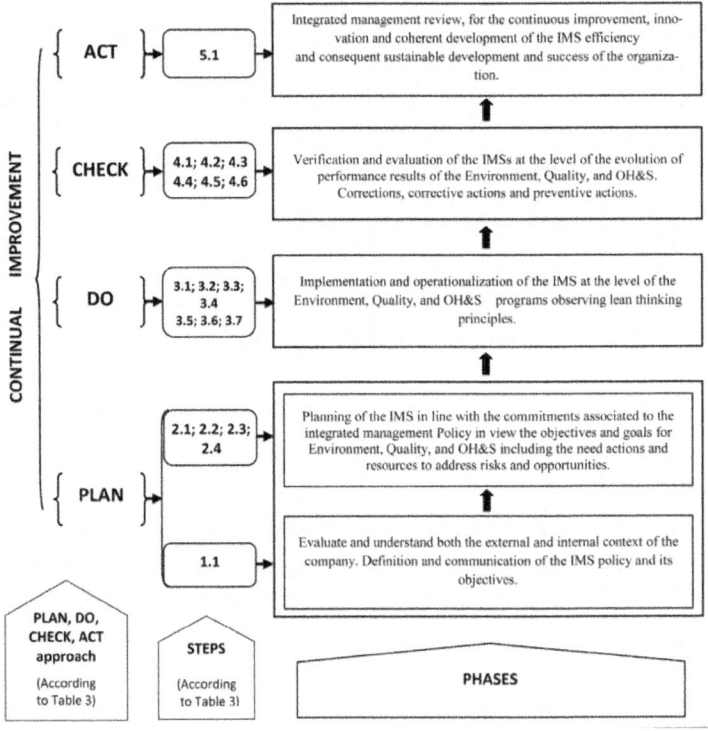

Figure 3: Model of development of the IMS Source: Adapted and upgraded from [11] [12].

In the case of ISO 9001 [3] it should be consid- ered associated with product realization, other complementary clauses, particularly in the context of resource al- location (6.1, 6.2, 6.3, 6.4) and management commitment (5.1, 5.5.1). Different activities of the organization should be set out as processes and put into operation, as per the applicable requirements, in order to strictly comply with the policies and instructions, whether they are documented or not, such as, to ensure that the or- ganization's objectives and targets are achieved and the different Stakeholders are satisfied [14] .

In the Phase III—CHECK, are considered six steps (4.1 to 4.6) designed to meet requirements of clauses: 4.5—Checking of the ISO 14001 [2] , 8—Measurement, analysis and improvement of ISO 9001 [3] , and 4.5—Checking of OHSAS 18001 [4] . With the exception of step 4.3—Investigation of incidents resulting from a specific sub-section, the 4.5.3.1—Incident investigation, the OHSAS 18001 [4] has no correspondence in the ISO 9001 [3] and ISO 14001 [2] . Critical information in the aim of the IMS should be identified, collected and analysed. Consistent with the integrated policy and the commitment to compliance, organization should estab- lish, implement and maintain documented records to demonstrate the improvements and compliance with all the applicable internal and external requirements. The use of KPIs—key process indicators, to monitor the processes, their control and continual improvement should be made systematically and guaranteed by the process owners, through the active involvement and participation of collaborators [14] . Particular relevance for the internal and external combined audits of the different components and areas of the IMS, and suppliers and subcontractors. Audits shall be scheduled at a frequency that takes into account the risk of the business. It must be conducted to assess the level of implementation and compliance of the IMS, its evolution, effectiveness and potentiate the identification of the necessary corrections and opportunities for improvement, which must be listed and priori- tized on its evaluation and implementation [14] [15] .

Finally, in the Phase IV—ACT, it was identified step 5.1—Critical analysis and review of the integrated management system, which refers to requirements of clauses: 4.6—Management review of ISO 14001 [2], 5.6— Management review of ISO 9001 [3] , and 4.6—Management review of OHSAS 18001 [4] . Top management should ensure that processes of assessment, improvement and innovation on the different components of the management system. Management reviews should be conducted at planned intervals, to identify opportunities for improvements based namely in the lean philosophy, to assess the need for modifications to the IMS and to the integrated management policy, to ensure that it continues to be appropriate, suitable, effective and efficient [14] . Top management should review the integrated management system, defining the review inputs according to the requirements of each MSSs. Management review requires, in itself, a very careful preparation phase, par- ticularly, to the level of various information that supports the inputs [11] [12] . Records of the management re- view should be retained [2] - [4].

DISCUSSION AND CONCLUSIONS

Management systems standards have developed in an unprecedented manner in the last few years. There exist at least, one for each interested part and new ones are going to be published. There is no international ISO standard with a specific structural model for Integrated Management Systems for Quality, Environment and Safety, or for other management areas such as: Risk Management; Information Security Management; RDI Management; and Social Responsibility Management, among others. The impact generated by environmental, quality, and safety and other MSSs is demonstrated by the importance of such standards worldwide, ISO 14001 [2] , ISO 9001 [3] , and OSHAS 18001 [4] . Organizations have to understand its context, the needs and expectations of interested parties and associated requirements, and consequently to determine the scope of the integrated management sys- tem, and formalise a policy to be communicated to internal and external interested parties.

The continuous improvement of global performance of organizations must be always a present goal in a per- spective of sustainability [43] . The development methodology to integrate various sub-management systems supported on a model of integration of the EMS, QMS, and OH & SMS at organizations should therefore poten- tiate, for each phase: Plan, Do, Check, Act, a careful and methodical analysis of the differences that effectively are observed at the level of normative requirements under the equivalent clauses and for each step of their de- velopment as the advocated model of integration.

Making MSSs requirements compatible through the analysis of their similarities promotes integration and can be depicted from the compatibility matrix presented—ISO14001 [2] , ISO 9001 [3] , and OHSAS 18001 [4] , es- tablishing correspondences, matching them with each other and associate the following phases of the PDCA cy- cle—Plan, Do, Check, Act: Policy and principles; Planning, Implementation and Operation, Performance Eva- luation, Improvement, Management Review. This is one of the activities that in the aim and objectives of the in- tegration model was given special attention in context of characterization and framework of the situation diag- nosed in the company in which the research was conducted. That compatibilization constitutes, the starting point for subsequent activities of integration, simplification and optimization, to a level of the strictly necessary and consequent integration maximized as desired of the three individual MSs—EMS, QMS, and OH & SMS, in con- text of strong competitiveness and changing of the business environments, which occurs at an accelerated and turbulent manner, enhancing also the sustained development of the business with added value for the relevant interested parts.

The integration of the three individual MSs represent added value both in the present and, fundamentally, for the future, not only for the company, as well as for a whole range of interested parties [43] . Examples are also highlighted by the surveyed respondents: the elimination of conflicts between individual systems with optimiza- tion of Resources; the improvement at the level of the coordinated and integrated management of the risk asso- ciated to the safety of the persons and company assets, environment and quality of the products; the reduction on the number of internal and/or external audits and audits to suppliers, and spent time versus associated costs; the creation of added value for the business through the elimination of several types of waste.

From the statistical analyses, resulting from the responses to the survey, there are shown a set of conclusions that by them self reveal: the importance, presently and for the future, of various "motivating factors" that were evaluated and alone justify and validate the model of implementation of the integration of the EMS, QMS, and OH & SMS in the company, either from internal aspects such as rationalization and optimization of resources, reduction of costs and bureaucracy; otherwise from external aspects, such as increasing competitiveness, to sat- isfy the growing demands of customers and others stakeholders. There were also identified a relevant number of difficulties, as well as a range of potential benefits resulting from the integration of the EMS, QMS, and OH & SMS into an IMS.

One of the major problems that organizations are facing with the integration of several MSSs is regarding the conception of a methodology and the implementation of an adequate structure of IMS to overcome the problems resulting from multiple MSSs. The continuous improvement of the global performance of organizations must be always a present goal in a perspective of sustainability and the route that should be taken by maximizing the in- tegration of the several individual MSs [13] , supported in an model of IMS, flexible, integrator and lean [15].

ACKNOWLEDGEMENTS

This work had the financial support of the Portuguese Foundation for the Science and Technology (FCT) through the Strategic Project-UI 4005-2014, Project Reference PEst-OE/EME/UI4005/2014.

REFERENCES

1.   International Organization for Standardization (ISO) (2008) The Integrated Use of Management System Standards. In- ternational Organization for Standardization, Geneva.

2. International Organization for Standardization (ISO) (2004) ISO 14001: Environmental Management Systems. Re- quirements with Guidance for Use. 2nd Edition, ISO Copyright Office, Geneva.

3. International Organization for Standardization (ISO) (2008) ISO 9001: Quality Management Systems—Requirements. 4th Edition, ISO Copyright Office, Geneva.

4. British Standards Institution (BSI) (2007) BS OHSAS 18001: Occupational Health and Safety Management Systems—Requirements. 2nd Edition, BSI Limited, London.

5. Karapetrovic, S. and Casadesus, M. (2009) Implementing Environmental with Other Standardized Management Sys- tems: Scope, Sequence, Time, and Integration. Journal of Cleaner Production, 17, 533-540. http://dx.doi.org/10.1016/j.jclepro.2008.09.006

6. Disterer, G. (2013) ISO/IEC 27000, 27001 and 27002 for Information Security Management. Journal of Information Security, 4, 92-100. http://dx.doi.org/10.4236/jis.2013.42011

7. International Organization for Standardization (ISO) (2009) ISO 9004: Managing for the Sustained Success of an Or- ganization—A Quality Management Approach. 3rd Edition, ISO Copyright Office, Geneva.

8. Bernardo, M., Casadesus, M., Karapetrovic, S. and Heras, I. (2009) How Integrated Are Environmental, Quality, and Other Standardized Management Systems? An Empirical Study. Journal of Cleaner Production, 17, 742-750. http://dx.doi.org/10.1016/j.jclepro.2008.11.003

9. Rebelo, M.F., Santos, G. and Silva, R. (2014) Integration of Individualized Management Systems (MSs) as an Aggre- gating Factor of Sustainable Value for Organizations: An Overview through a Review of the Literature. Journal of Modern Accounting and Auditing, 10, 357-384.

10. Khanna, H.S., Laroiya, S.C. and Sharma, D.D. (2010) Integrated Management Systems in Indian Manufacturing Or- ganizations: Some Key Findings from an Empirical Study. The TQM Journal, 22, 670-686. http://dx.doi.org/10.1108/17542731011085339

11. Rebelo, M.F. (2011) Contribution to the Structuring of a Model of Integrated Management System QES. Master Thesis, Polytechnic Institute of Cávado and Ave, Barcelos.

12. Santos, G., Rebelo, M.F., Barros, S. and Pereira, M. (2012) Certification and Integration of Environment with Quality and Safety—A Path to Sustained Success. In: Curkovic, S., Ed., Sustainable Development—Authoritative and Leading Edge Content for Environmental Management, InTech, Croatia, 193-218. http://dx.doi.org/10.5772/48414

13. Rebelo, M.F. and Silva, R.G. (2012) Integration of Individual Management Systems—An Organizational Pillar for the Competitiveness and the Sustainability of Business. The International Conference on Innovation for Sustainability—IS 2012, Porto, 27-28 September 2012, 1-12.

14. Rebelo, M.F., Santos, G. and Silva, R. (2014) A Generic Model for Integration of Quality, Environment and Safety Management Systems. TQM Journal, 26, 143-159.http://dx.doi.org/10.1108/TQM-08-2012-0055

15. Rebelo, M.F., Santos, G. and Silva, R. (2014) Conception of a Flexible Integrator and Lean Model for Integrated Management Systems. Total Quality Management & Business Excellence, 25, 683-701. http://dx.doi.org/10.1080/14783363.2013.835616

16. Simon, A., Bernardo, M., Karapetrovic, S. and Casadesus, M. (2011) Integration of Standardized Environmental and Quality Management Systems Audits. Journal of Cleaner Production, 19, 2057-2065. http://dx.doi.org/10.1016/j.jclepro.2011.06.028

17. Olaru, M., Maier, D., Nicoara, D. and Maier, A. (2014) Establishing the Basis for Development of an Organization by Adopting the Integrated Management Systems: Comparative Study of Various Models and Concepts of Integration. Procedia-Social and Behavioral Sciences, 109, 693-697. http://dx.doi.org/10.1016/j.sbspro.2013.12.531

18. Asif, M., Bruijn, E.J., Fisscher, O.A.M. and Searcy, C. (2010) Meta-Management of Integration of Management Sys-tems. The TQM Journal, 22, 570-582.http://dx.doi.org/10.1108/17542731011085285

19. Nowicki, P. (2013) Risk Management—An Important Issue in Quality Management Systems. The 7th International Quality Conference, Kragujevac, 24 May 2013, 267-272.

20. Suditu, C. (2007) Positive and Negative Aspects Regarding the Implementation of an Integrated Quality-Environ- mental-Health, and Safety Management System. Annals of the Oradea University, Fascicle of Management and Tech- nological Engineering, 6, 2013-2017.

21. De Oliveira, O.J. (2013) Guidelines for the Integration of Certifiable Management Systems in Industrial Companies. Journal of Cleaner Production, 57, 124-133.http://dx.doi.org/10.1016/j.jclepro.2013.06.037

22. Simon, A., Bernardo, M., Karapetrovic, S. and Casadesus, M. (2013) Implementing Integrated Management Systems in Chemical Firms. Total Quality Management and Business Excellence, 24, 294-309. http://dx.doi.org/10.1080/14783363.2012.669560

23. Santos, G., Mendes, F. and Barbosa, J. (2011) Certification and Integration of Management Systems: The Experience of Portuguese Small and Medium Enterprises. Journal of Cleaner Production, 19, 1965-1974. http://dx.doi.org/10.1016/j.jclepro.2011.06.017

24. Vinodkumar, M.N. and Bhasi, M. (2011) A Study on the Impact of Management System Certification on Safety Management. Safety Science, 49, 498-507.http://dx.doi.org/10.1016/j.ssci.2010.11.009

25. Esaki, K. (2013) General Frame Work of New TQM Based on the ISO/IEC25000 Series of Standard. Intelligent Information Management, 5, 126-135.http://dx.doi.org/10.4236/iim.2013.54013

26. Zeng, S.X., Xie, X.M., Tam, C.M. and Shen, L.Y. (2011) An Empirical Examination of Benefits from Implementing Integrated Management Systems (IMS). Total Quality Management and Business Excellence, 22, 173-186.http://dx.doi.org/10.1080/14783363.2010.530797

27. Alolayan, S., Hashmi, S., Yilbas, B. and Hamdy, H. (2013) An Empirical Evaluation of the ISO 9001 Quality Man- agement Systems for Certified Work Organizations in Kuwait as Benchmarked against Analogous Swedish Organiza- tions. Journal of Service Science and Management, 6, 80-95. http://dx.doi.org/10.4236/jssm.2013.61009

28. Daub, C.H. and Ergenzinger, R. (2005) Enabling Sustainable Management through a New Multi-Disciplinary Concept of Customer Satisfaction. European Journal of Marketing, 39, 998-1012. http://dx.doi.org/10.1108/03090560510610680

29. Tsai, W.H. and Chou, W.C. (2009) Selecting Management Systems for Sustainable Development in SMEs: A Novel Hybrid Model Based on DEMATEL, ANP, and ZOGP. Expert Systems with Applications, 36, 1444-1458.http://dx.doi.org/10.1016/j.eswa.2007.11.058

30. Salomone, R. (2008) Integrated Management Systems: Experiences in Italian Organizations. Journal of Cleaner Production, 16, 1786-1806. http://dx.doi.org/10.1016/j.jclepro.2007.12.003

31. Granerud, L. and Rocha, R.S. (2011) Organisational Learning and Continuous Improvement of Health and Safety in Certified Manufacturers. Safety Science, 49, 1030-1039. http://dx.doi.org/10.1016/j.ssci.2011.01.009

32. Karapetrovic, S. and Jonker, J. (2003) Integration of Standardized Management Systems: Searching for a Recipe and Ingredients. Total Quality Management & Business Excellence, 14, 451-459. http://dx.doi.org/10.1080/1478336032000047264

33. Zeng, S.X., Shi, J.J. and Lou, G.X. (2007) A Synergetic Model for Implementing an Integrated Management System: An Empirical Study in China. Journal of Cleaner Production, 15, 1760-1767. http://dx.doi. org/10.1016/j.jclepro.2006.03.007

34. Santos, G., Barros, S., Mendes, F. and Lopes, N. (2013) The Main Benefits Associated with Health and Safety Man- agement Systems Certification in Portuguese Small and Medium Enterprises Post Quality Management System Certi- fication. Safety Science, 51, 29-36.http:// dx.doi.org/10.1016/j.ssci.2012.06.014

35. Cempalavras, Comunicação Empresarial (2012) Portuguese Certified Companies Guide. 7th Edition. http://www.cempalavras.pt/GEC_2012/ EN/index.html

36. Cempalavras, Comunicação Empresarial (2013) Portuguese Certified Companies Guide. 8th Edition. http://www.cempalavras.pt/GEC_2013/ EN/index.html

37. Jørgensen, T.H. (2008) Towards More Sustainable Management Systems: Through Life Cycle Management and Integration. Journal of Cleaner Production, 16, 1071-1080.http://dx.doi.org/10.1016/j. jclepro.2007.06.006

38. Almeida, J., Sampaio, P. and Santos, G. (2012) Integrated Management Systems—Quality, Environment, and Health and Safety: Motivations, Benefits, Difficulties, and Critical Success Factors. Book of Abstracts, The International Symposium on Occupational Safety and Hygiene— SHO 2012, Guimarães, 9-10 March 2012, 13-15.

39. Almeida, J., Domingues, P. and Sampaio, P. (2014) Different Perspectives on Management Systems Integration. Total Quality Management and Business Excellence, 25, 338-351.http://dx.doi.org/10.1080/14783363.2 013.867098

40. International Organization for Standardization/International Electrotechnical Commission [ISO/IEC] (2005) ISO/IEC 17025: General Requirements for the Competence of Testing and Calibration Laboratories. ISO Copyright Office, Geneva.

41. Instituto Português da Qualidade (2009) NP 4463: Linhas de orientação sobre técnicas estatísticas para a ISO 9001: 2000 (ISO/TR 10017:2003). Instituto Português da Qualidade, Caparica.

42. Saeed, B. and Wang, W. (2014) Sustainability Embedded Organizational Diagnostic Model. Modern Economy, 5, 424- 431. http://dx.doi. org/10.4236/me.2014.54041

43. Rebelo, M.F. and Santos, G. (2012) Integration of the Occupational Health and Safety Management System with the Quality Management System and Environmental Management System—From the Theory to the Action. Book of Abstracts, The International Symposium on Occupational Safety and Hygiene—SHO 2012, Guimarães, 9-10 March 2012, 372-374.

Chapter 7

GEOLOGICAL AND GEOTOURISM STUDY OF IRAN GEOLOGY NATURAL MUSEUM, HORMOZ ISLAND

Abdollah Yazdi[1], Mohammad Ali Arian[2], Mahmoud M. Rezapour Tabari[3]

[1]Department of Geology, Kahnooj Branch, Islamic Azad University, Kerman, Iran

[2]Department of Geology, North Tehran Branch, Islamic Azad University, Tehran, Iran

[3]Department of Engineering, Shahrekord University, Shahrekord, Iran

ABSTRACT

Iran is a country that benefits from nice nature, diverse continent, areas full of unique geological phenomena. Thus, it is necessary to study these attractions for better recognition of them. In this regard, Hormoz Island with valuable geoheritage, biodiversity, cultural, historical and political diversity is very important. The accumulation of these attractions and its being located in Persian Gulf strategic area made it of considerable significance in national and international communities. Hormoz Island is a spherical salt dome which is located in Hormoz strait. This Island is composed of evaporites, igneous rocks and sedimentary rocks, and sediments mainly belong to Mishan and Aghajari formations and salt, gypsum and, to a lesser extent, limestone evaporites. In sedimentary formations of Hormoz, Aouthigenic minerals such as Pyrite, Dolomite, quartz, Anhydrite, gypsum and halite are frequently seen. Mineralization of volcanic leads to formation of high temperature minerals, such as oligiste, pyroxene, amphibole and low temperature and hydrothermal minerals such as Pyrite, quartz... in the tracks of igneous rocks. This unique geodiversity in rocks and mineral which made various colors in Hormoz Island made it a mineraogical reservoir. Beside considerable mineralogical attractions, there are other potentials like ochre mine, coral reefs, rock seashore, sea caves, salt caves, plant cover and wild life which have added to various tourism capabilities of this Island and made it a unique place in the world. This paper studies Hormoz Island in terms of geological features and geotourism potentials.

INTRODUCTION

In recent years, tourism industry has become one of the main poles of development from various respects, especially economic source. It is such that most authorities believe that tourism will become dominant industry in near future and will have various socioeconomic effects (Bayati Khatibi, 2010) [1] . Due to job creation and relatively rapid profiting characteristic, tourism is a proper ground for foreign investment and can accelerate tourism development, promote its economic criteria and bring out new ideas, technologies and markets (Papeli Yazdi, 2011) [2] . Geotourism, as a subcategory of tourism, is considered one of the new methods in providing tourism attraction (Servati, 2008) [3] and has allocated a main part of tourism studies to itself. Iran has a nice nature, diverse continent and areas full of unique geological phenomenon so that it seems necessary to investigate these attractions for their better cognition and geotourism development (Yazdi, 2012) [4] .

Iran is one of the few countries of the world with beautiful natural and geological phenomenon due to its unique and excellent geographical condition. In this regard, Hormoz Island has become geological paradise due to certain geological condition and unique diversity of minerals and rocks and attracted many geotourists. This spherical island has been located in Persian Gulf' entrance in 56°25' to 56°30' east longitude and 27°2' to 27°6' north altitude. Its long diameter is maximumly 9 km, its short diameter is almost 5.5 km, and it covers an area about 41 km (Figure 1). This Island is an adult salt dome and its morphological, mineralogical and lithological diversity is significant. Besides geological attractions, this area benefits from historical attractions like Portuguese castle and natural attractions such as alga seashore, wildlife, etc. which will be investigated in this paper.

METHODOLOGY

This research is practical and developmental and the methodology used in it is descriptive-analytical. Based on this, library-documentary studies, interpretation and analysis of satellite images, various field studies and direct observation of phenomenon have been used.

GEOTOURISM

Geotourism is a relatively new concept in tourism industry which has considerable growth in recent decade. Geotourism has a certain definition with geological tourism at its centre (Newsome & Dowling, 2006) [5] and deals with the investigation of related forms and consequences to earth, geomorphologic and geological phenomena. According to Gates (2006) [6] , geotourism means "tourism in geological outlooks". Geotourism, according to Dowling & Newsome (2006) [7] , deals with geology, geomorphology, natural outlooks and the forms of earth surface, layers with fossil, rocks and minerals with emphasis on the creating processes. Furthermore, it can be argued that geotourism is informed and responsible tourism in nature with the aim of looking at recognition of geological phenomena and processes and learning their formation and revolution (Amrikazemi, 2009) [8] . According to the above definitions, geotourism is not only is new part of tourism market, it is a principal guidance to help maintaining nature and sustainable development, which is compatible with the economic equilibrium, social condition and ecology and complements them.

GEOLOGY OF HORMOZ ISLAND

This Island is one of the biggest and most famous salt domes of Persian Gulf whose formation has begun from

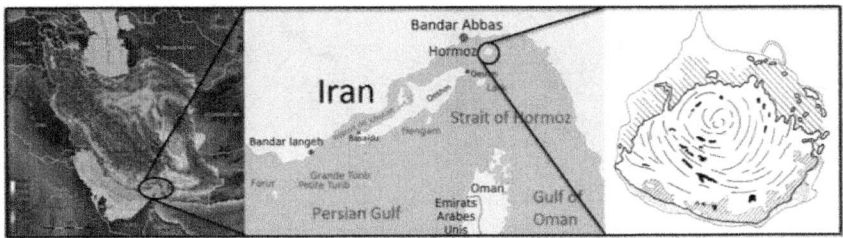

Figure 1: The location of Hormoz Island in Iran.

late Precambrian and come out of water about 50,000 years ago. Hormoz Island is almost structurally concentric with evaporites, igneous rocks and sedimentary rocks layers. Bland Ford in 1872 called the collection of igneous, metamorphic and evaporites rocks of this area as Hormoz Series.

According to Elyasi et al. (1975) [9] , this Island is composed of evaporites rocks, ferroginous formations and Miocene-Pliocene sediments (Figure 2). Beside evaporites, sedimentary rocks and igneous are can be find which has been previosly called Hormoz formation. Stocklin [10] in his study from 1961 to 1968 concluded that expression Hormoz series is true for underlying salt layers and gypsum and sandy rocks and limestone with fossil can be separated and limit the expansion of evaporites in late Precambrian in northern-southern extent of wide area which reaches Oman-Naiband fracture from east and Qatar-Kazeron fracture from west (Stocklin, 1972) [10] . In overall, Hormoz series collection can be attributed to a row of evaporites, igneous and rarely metamorphic rocks which have been deposited after late Precambrian. These rocks began with a complete cycle with dark and usually dolomite carbonate rocks and sediment on chalk, salt and shale. Concerning the similarity of Hormoz series fossils and fossils on lower part of Valdai series in Russia platform and late part of Precambrian in Australia, Hormoz series can be considered as belonging to late Precambrian.

Various magma activities in Magma Island are seen in form of extrusive and intrusive. Igneous events happened in two phase, the first one includes analysed basalt and diabase and the second one includes rhyolite, rhyodacite and trachyte which respectively belong to Permian and Triassic. The mentioned rocks, except rhyolite (Figure 3) which is somehow intact, are mainly decomposed.

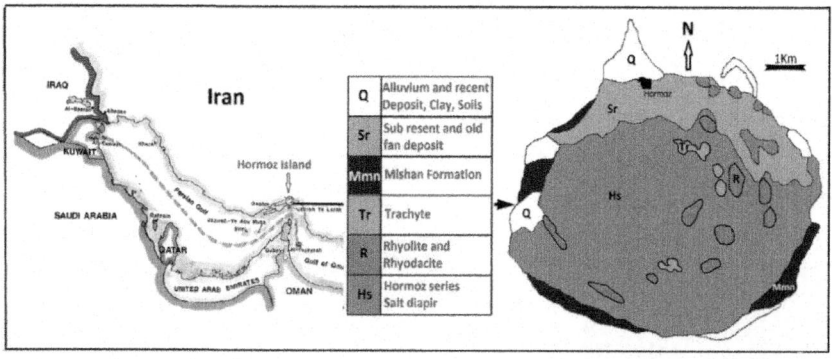

Figure 2: Geology map of Hormoz Island by corrections (Elyasi et al., 1975) [9] .

Figure 3: An overview of Trakyt-Riolite mass of Hormoz Island.

HORMOZ SLAT DOME

In southern Zagros area, there is about 130 salt domes which erosion has created beautiful perspectives in them (Figure 4) such as salt cave in these salt domes. In lower layers of Persian Gulf area, there is a thick bed of salt in 4 - 10 km depth which due to special weight less than lateral rocks and the weight pressure of the upper layers, penetrated in higher layers and reached surface and created various domes. This salt movement might happen in faults through sedimentary layers... The width of most salt domes is about 1 to 10 km and their big dome-like mountains are usually salt lump however, their surface is usually covered by red soils.

The main soil of its covering rock with several meter thickness is in fact the residual of insolvable materials from the main salt beds. The dominant soil of this bed is clay and silt with usually more than 50% plaster-rock. The red colour is up to 15% related to iron oxide that is usually in form of hematite. Some black salt domes are frequently seen. The lump values for salt domes in southern Zagros are usually 2 to 6 mm. Although it seems that in some sample, 15 mm growth has happened (Waltham, 2008) [11] .

Hormoz salt dome has come out of water about 50,000 years ago and due to severity of salt tectonic movements, Neogenes sedimentary layers reached 75-degree slope. Raining has been able to be effective in salt domes in creation of cavities and salt caves as can be seem in seashore salt domes "Namakdan and Hormoz" in Qeshm and Hormoz Islands which have been identified and drawn by geologists from Czech (Figure 5).

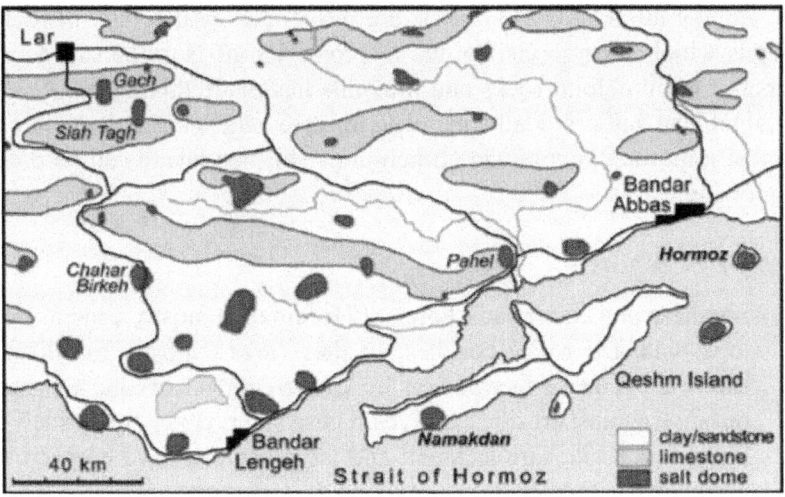

Figure 4: The situation of salt domes of south Zagros and Persian Gulf.

Figure 5: Layering of salt and iron oxide in roof of salt cave of Hormoz salt dome.

GEOLOGICAL AND GEOTOURISM ATTRACTION OF HORMOZ ISLAND

In addition to attractive biological outlooks of Hormoz Island, the geological attractions and perspectives of this Island is very interesting and extraordinary. Eye-catching attractions of Hormoz are influenced by diversity of rocks and minerals which became visible due to protrusion of Hormoz salt done. The diversity of soil colour, rocks and minerals has made this Island a colourful natural tableau and provided a paradise for researchers of geology. In overall, the most important geotourism attraction of Hormoz Island can be presented as following.

Hormoz Seashores

- Southern and eastern seashores of Hormoz are mostly constituted from low, high and rocky beaches. In these areas, mostly fossilized lime layers are at the surface and on the less-strength layers and are relatively resistive against erosion of waves. These layers cover non-resistive units like cover rock. Various small and large cavities have been created in lower parts of rocks due to collision of waves and erosion. These dents have been formed at the bottom of rocks and would become larger over time (Figure 6 & Figure 7).

- Another part of Island seashore near to ocher (red soil) mine is known as red seashore. Due to collision of seawater and this soil which has high iron oxide, the colour of this beach turns red and creates a beautiful scene (Figure 8).

- Other parts of Hormoz Island are known as colourful beaches. Each area of its colourful mountain represents a beautiful and extraordinary beach and has unique diversity (Figure 9).

Figure 6. Precipice cliffs and rocks of eastern south of Hormoz (Aghajari formation has creamy colour and Mishan formation is green in farther areas).

Figure 7. A nice view of southern seashores of Hormoz Island.

Figure 8. Red seashore around Hormoz ochre/red soil mine.

Figure 9: Colourful mountainous seashores of Hormoz Island.

- Other parts of Hormoz beach is constituted from beautiful ripple mark covered with oligiste crystals which make this area shiny especially at the time of sunshine (Figure 10).

Red Soil Mine (Ochre Mine)

Currently, this mineral material has been seen in Persian Gulf area in salt domes and some Islands. Among various mines in this area, Hormoz ochre mine can be referred to which is the most significant one. This mine has a reservoir about 390,000 tons and is unique in terms of quality and application in industries all over the world (Figure 11). The soil of this mine has application in some industries such as cosmetics, paint and rust, colourful glasses, paper, enamel, ceramic, rubber, colouring construction materials, chemical fertilizer and...

(Aqanabati, 2006) [12] . Some years ago the soil of this mine was exported to countries such as Britain, France, Portuguese, India, EAU, and Pakistan and so on, however, now the export of this mineral material is controlled and it is less exported.

This mineral material is seen in form of lens shape masses and has been created due to washing of ferrous rocks by penetrating waters and transfer of feo^{+3} to the earth. Hydration of ferrite rick leads to creation of different kinds of hydrated oxides like limonite and goethite (Moeinvaziri, 1996) [13] which have created yellow and red mountains (especially in Qeshm-Hormoz path).

Native habitants of this Island call the red soil of this area as "Gelak" and use it in different kinds of food as spice. One of these foods is "Suragh" delicious food which is prepared by soaking Sardine fish in Hormoz red soil and orange and is served with bread (Figure 12).

The Diverse Minerals of Hormoz Island

Mineralization processes in igneous rocks lead to creation of various beautiful minerals such as oligiste, pyroxene, amphibole, pyrite, quartz, etc. in tacks of igneous rocks. This unique diversity in rocks and minerals has

Figure 10: Ripple marks of seashore covered with oligiste mineral.

Figure 11: A layout of ochre mine of Hormoz Island.

Figure 12: The area where native habitants of the region use its soil for food.

created various colours in Hormoz Island and made it a treasure of minerals (Yazdi, 2013) [14] .

Gypsum: These crystals can be found in some parts of Islands in form of rhombohedral. These crystals will turn to anhydrite white powder due to dehydration (Figure 13).

Pyroxene: In southern west of Hormoz Island, pyroxene has been formed between tracks of rhyolite masses. These crystals can be seen in 5 cm length along with oligiste and alpha quartz (Moeinvaziri, 1996) [13] , (Figure 14).

Oligiste: High temperature oligiste crystals can be seen in form of bipyramid, hexagonal or in form of simple hexagonal pyramid, low temperature oligiste crystals can be seen in form of thin hexagonal aglet with rhombohedral surfaces in igneous rocks or in sedimentary rocks (shale and dolomite) (Figure 15).

Pyrite: This mineral material has been created in form of didodecahedral and octahedral in Hormoz igneous tuffs by fumerole. In addition to tuffs, pyrite automorph crystals can be seen in sedimentary rocks (Figure 16).

Apatite: Apatite crystals are frequent in ferrous riolite breccia and due to high frequency of these crystals, even they can be thought as a source for rare earth elements (Figure 17).

Dolomite: Dolomite is seen in form of big and automorph crystals in form of grayish and white rhombohedral and pinacoidal in sediments. Small and automorph crystals can be seen in some cavities (Figure 18).

Quartz: It is seen in form of hexagonal pyramids with rhombohedral surface.

Salt Diverse Structure in Hormoz Island

Since Hormoz Island is a salt dome, considerable structures of various salt forms can be seen in central parts of Island (Figure 19).

Figure 13: Gypsum beautiful crystals.

Figure 14: Pyroxene crystals of Hormoz Island.

Figure 15: Wrinkled layers containing oligiste.

Figure 16: Pyrite crystals.

Figure 17: A sample Apatite in Hormoz Island.

Colourful Soils of Hormoz Island

One of the amazing and wonderful geological phenomena of Iran is colourful mountains of Hormoz Island which have created landscape attractions of universe (Figure 20). It is an Island which is natural for geologists and an original scene for artists. What distinguishes Hormoz from other Islands of Persian Gulf is its colourful rainbow soil which has been formed due to diverse rocks and minerals. Thus, most artists design various shapes or carpets with this colourful soil which is unique and outstanding (Figure 21).

Figure 18: A sample Dolomite in Hormoz Island.

Figure 19: Diverse samples of solved salts and gypsum in Hormoz Island.

Figure 20: A view of unique diversity of coloured mountains of Hormoz.

Figure 21: Carpets made of coloured soils of Hormoz.

CONCLUSION

What was mentioned in this paper indicates that considering the global significance of geotourism industry, Hormoz Island which is one of the most beautiful and important salt domes of universe, has high potential for geotourism development. Diversity in geological perspectives and numerous minerals of Hormoz Island, which made it a natural museum for fonder of geology, is a strong potential for making this region be one of the world geotourism poles and a proper area for researchers and scholars. In this way, researchers are able to take proper actions for scientific development of geotourism in this area with stronger foundations obtained from research finding and transfer scientific grounds of phenomenon and geo-sites to visitors. In this condition, geoconservation (conservation of geological phenomena) has been properly performed by tourists. Moreover, geotourism finds its real meaning besides economic development, thus, Hormoz Island can be a good pattern for achieving the above objectives.

REFERENCES

1. Bayati Khatibi, M., Shahabi, H. and Qaderi Zadeh, H. (2010) Geotourism: A New Approach in Utilization of Geomorphologic Attractions. Journal of Geographical Space, 29, 27-50.

2. Papeli Yazdi, M.H. and Saghaei, M. (2011) Tourism, Nature & Concepts. 6th Edition, Samt Publication, Tehran, 6.

3. Servati, M.R. and Qasemi, A. (2008) Geotourism Strategies in Fars. Journal of Geographical Space, 2, 6.

4. Yazdi, A. (2012) A Study of Iran's Lut Desert: Geomorphological and Geotourism Attractions. Proceedings of Annual International Conference on Geological & Earth Sciences (GEOS2012), Singapore, 3-4 December 2012, 35-41.

5. Newsome, D. and Dowling, R.K. (2006) The Scope and Nature of Geotourism. In: Dowling, R.K. and Newsome, D., Eds., Geotourism, Chapter One, Elsevier, Oxford, 3-25.

6. Gates, A.E. (2006) Geotourism: A Perspective from the USA. In: Dowling, R.K. and Newsome, D., Eds., Geotourism, Chapter Nine, Elsevier, Oxford, 157-179.http://dx.doi.org/10.1016/B978-0-7506-6215-4.50017-8

7. Dowling, R.K. and Newsome, D. (2006) Geotourism's Issues and Challenges, Geotourism, Chapter Thirteen, Elsevier, Oxford, 242-254.

8.   Amrikazemi, A. (2009) Atlas of Geopark & Geotourism Resources of Iran. Geological Survey of Iran Publication, Tehran, 22-23.

9.   Elyasi, J., Aminsobhani, E., Behzad, A., Moeinvaziri, H. and Meysami, A. (1975) Geology of Hormoz Island. Geological Survey of Iran Publication, Tehran, 1, 13.

10. Stocklin, J. (1972) Lexique Stratigraphique International. Vol. 3, Geological Survey of Iran Publication, Tehran, 6, 15.

11. Waltham, T. (2008) Salt Terrains of Iran. Geology Today, 24, 188-194. http://dx.doi.org/10.1111/j.1365-2451.2008.00686.x

12. Aqanabati, S.A. (2006) Geology and Mineral Potential of Hormozgan Province. Journal of Development of Geology Training, 12, 4-11.

13. Moeinvaziri, H. (1996) An Introduction of Magmatism in Iran. University of Tarbiat Moalem Publication, Tehran, 440.

14. Yazdi, A., Emami, M.H. and Jafari, H.R. (2013) IRAN, the Center of Geotourism Potentials. Journal of Basic and Applied Scientific Research, 3, 458-465.

Chapter 8

HYDROGEOLOGICAL FRAMEWORK AND GROUNDWATER BALANCE OF A SEMI-ARID AQUIFER, A CASE STUDY FROM IRAN

Leila Khodapanah[1], Wan Nor Azmin Sulaiman[1], Hamid Reza Nassery[2]

[1]Department of Environmental Sciences, Faculty of Environmental Studies, Universiti Putra Malaysia, Serdang, Malaysia

[2]Faculty of Earth Science, Shahid Beheshti University, Tehran, Iran

ABSTRACT

Climate changing and associated factors combined with considerably increases in water demand have been accompanied by severe depletion of reservoir storage of the most groundwater supplies of Iran. Shahriar aquifer in west of Tehran is a representative aquifer of these kinds. In order to meet water demand of the area and protecting groundwater from quantity and quality deterioration, precision recognition of geology, hydrologic and hydrogeologic characteristics of the aquifer is first step. The basic objective of this study is to develop the hydrogeological framework of the groundwater system in Shariar, Iran and to estimate groundwater balance as a scientific database for future water resources delevopment programs. Based on this research lateral groundwater inflows, direct infiltration of rainfall, stream bed infiltration, irrigation return and surplus drinking and industrial water are the recharging factors of the aquifer. Subsurface outflows, domestic and industrial pumping wells and agricultural abstraction are the main parameters discharge the aquifer system. Water balance in the Shahriar aquifer system is in disequilibrium and a deficit of about 24.7 million cubic meters exists.

INTRODUCTION

Development of hydrogeological investigation with emphasis on recharge rate estimation is one of the basic steps for a suitable water management [1]. This constitutes a major issue in regions with large demands for ground water supplies, such as in semiarid areas, where such resources are the key to agricultural and industrial development [2-4]. There are various methods for aquifer recharge estimations. Physical and chemical methods are the two main categories for estimating aquifer recharge rate [5,6]. Water table fluctuation is one of the most applied methods in recharge estimation. This method calculates the ground water storage change with considering water table fluctuations and the storage parameter. This method is considered to be one of the most promising and attractive due to its accuracy, ease of use and low cost of application in semiarid area [7]. The water table fluctuation method was first used to estimate ground water recharge and has since then has been used in numerous studies for the same purpose [8].

Shahriar plain, situated in west of Tehran and between two rivers Kan and Karaj, was someday considered as one of the main centers of producing sapling and seed in the country. In 1960, Amir Kabir multipurpose dam was constructed on Karaj river for flood storage and regulation, supplying drinking water for Tehran, capital of Iran, agriculture water regulation and supplying hydroelectric power [9]. After several years despite of covering forenamed aims, downstream lands including Shahriar plain was encountered water shortage. Lack of surface water in downstream and feasibility of drilling deep wells has deployed using of Shahriar's groundwater, as these supplies has become main water source of area [10]. Over-exploitation of groundwater resources as a consequent of population growth and rapid development in agriculture and industry has caused water scarcity and drop in groundwater level. Semiarid climate and not recharging aquifer through Karaj River has exacerbated the condition of Shahriar aquifer.

In view of this, there is an urgent need to reevaluate the status of water resources, quantity and quality monitoring of groundwater and better recognition of present aquifer condition for optimum utilizing of water resources in this area. The purpose of this study is to define a conceptual framework of the Shahriar aquifer, to consider the present groundwater resource potential and hydrodynamic characteristics of geological formation of study area. The discharge and recharging parameters of aquifer and water balance of Shahriar groundwater are also assessed.

MATERIALS AND METHODS

Study Area

The Shahriar plain covers an area of 580 km^2 and lies between longitudes 50° and 50°15' east and latitudes 35°30' and 35°45' north (Figure 1). The study area is bounded southern submontane of Central Alborz mountains to the north and conglomerate hills to the south, Karaj River to the west and Kan River to the east. The main towns in the area are Shahriar, Saeid Abad, Eslamshahr, Ghale Hasan Khan and Robat Karim. Topographic elevations range from 1020 to 1200 m above sea level. Mean plain elevation is 1100 m above sea level. The slope of study area in northeastern and eastern regions is toward south and in northwestern and western is toward southeastern. The general slope is roughly 0.7%.

Meteorological and Hydrologic Conditions

Prevailing climate of study area has semi arid characteristics. According to data of three meteorological stations

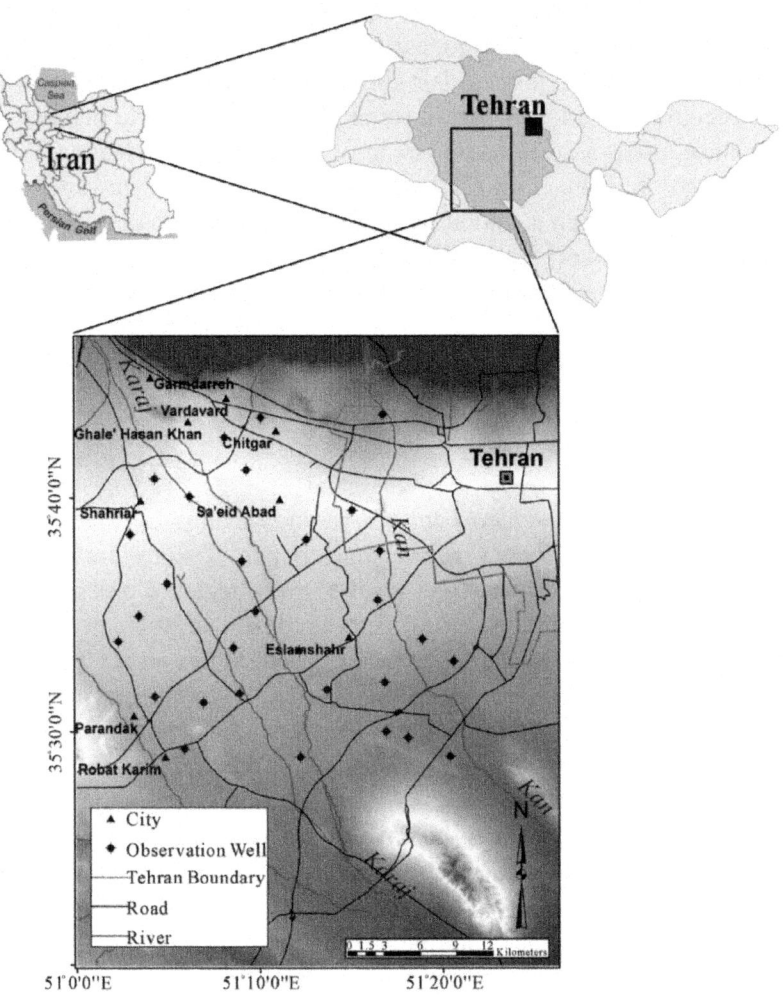

Figure 1: Study area map showing the location of monitoring boreholes.

existing in the area, average annual precipitation recorded for the years 1967-2005 is about 230 mm. The wet season begins from November and continues till May. The maximum rainfall occurs in February (15% - 18%) and winter is the most rainy season of the year. The average minimum temperature is 3.5°C in January and the average maximum is 30°C in August. Mean monthly relative humidity is about 67% in January and minimum relative humidity is 26% during the July-September period [11].

The average annual pan evaporation rate measured at Karim Abad station is about 2270 mm. According to Demarton and Emberger climate classification [12], the study area is dry and cold.

There are three rivers in northern mountains which include the Kan, Chitgar and Karaj from east to west. Kan and Karaj are perennial rivers. During wet season Kan joins to Jajrud river in south of Tehran, meanwhile the others become dry in lower part in south of the region due to percolation and evaporation losses as well as diversion of the water for irrigation. All rivers in the region originate from northern catchments and flow toward south and southeast. Three gauging station on main river are Bileghan station on Karaj river and Kahrizak and Sooleghan stations both on the Kan river. The mean discharges of these stations are 15.87, 0.8 and 2.46 m³ per second respectively (Figure 2). The mean discharge of Chitgar river is about 0.5 m³ per second which is calculated based on flow-duration relationships between basins with similar characterisics.

Geological Setting

Shahriar plain is located between Alborz and Central Iran tectonic zones. Present condition and topographic characteristics of study area in south of Central Alborz mountain has been resulted from tectonic processes and the three main river alluvial fans during time, as illustrated in the geological map (Figure 3). The study area is bounded between Alborz mountain and some small conglomerate hills to north and height of different lithology to the south. Central and southern smooth parts of plain form main aquifer and originate from the filling of a tectonic depression with quaternary alluvial. The oldest formation in the area outcrop on the Alborz mountain and southern height and are represented by tuff, andesitic and pyroclastic rocks (Eocene) [13,14]. These rocks which has deposited during the Laramian (laramid) orogenic phase are called Karaj Formation. In the Miocene epoch, there was a marine regression and a change to continental condition, mainly lacustrine, with the deposition of high-colored marls, gypsum, conglomerate and sandstone.

The sediments deposited in this area in the Pliocene epoch caused by Pasadenian orogeny phase are called Hezar Darreh formation. Due to resuming sedimentation till prior quaternary, these sediments are also known as Plio-Quaternar or Late Quaternary Alluvial [15]. Hezar Darreh formation, derived from the uplift of the Alborz Range on the north (Karaj formation tuff), consist of thick coarse light-colored conglomerates and sandstones

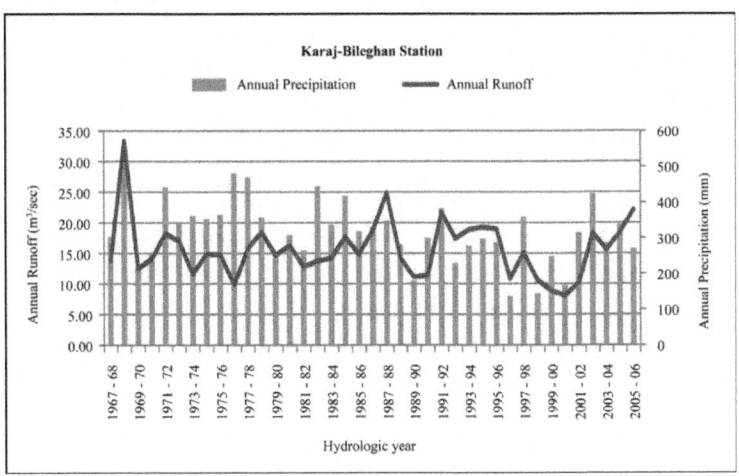

Figure 2: Annual precipitation and runoff in the Karaj-Bileghan station.

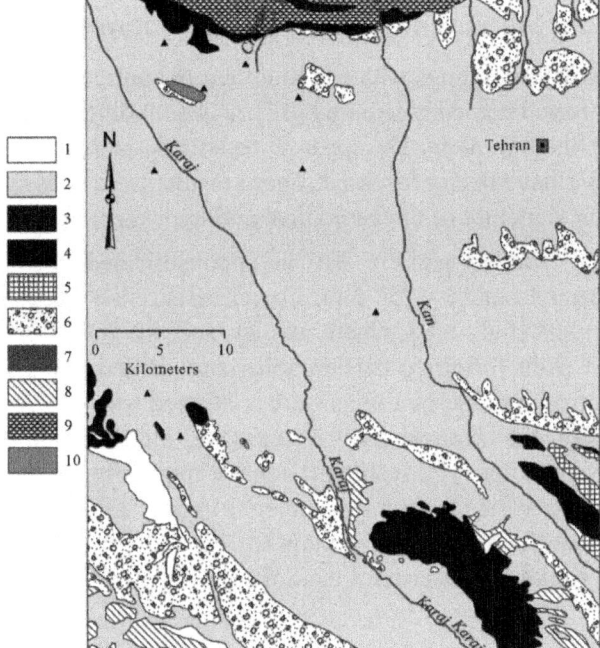

Figure 3: Geological map of the Shariar plain. 1) Alluvial terraces; 2) Alluvium (including TEHRAN ALLUVIUM); 3) Andesite/dacite lava and pyrocalstics marl; 4)

Andesitic lava and pyrocalstics; 5) Conglomerate (including HEZARDARREH FOR-
MATION; Mio-Pliocene); 6) Conglomerate (including KAHRIZAK FORMATION;
plio-pleistocene); 7) Mudstone, shale, dacite, pyroclastics; 8) Mudstone, gypsiferous;
9) Undifferentiated mudstone, shale, dacite, pyroclastics marl and andesite lava; 10)
Upper tuff member: andesite/dacite lava and pyroclastics, shale, marl, sandstone.

more than 1000 - 2000 meters thick near the mountains but increasing to
5000 meters in the eastern Tehran (type profile). This conglomerate alluvium
exposed in north of study area (Chitgar and between Vardavard and Garm
Darreh) and in southern part as single conglomerate hills.

Most of the Shahriar plain is built on an alluvial fan complex of Recent
and Quaternary age that thickens towards the south and thins towards the
Alborz Mountains. The Quaternary sediments can be divided to three groups:
Kahrizak alluvial formation, Tehran alluvial formation and Recent alluvials
[16]. Kahrizak deposits consist of silt and clay and outcrop in south of area
near Kahrizak and Mafi Abad. Tehran alluvial, derived from Hezar Darreh
erosion, consist of red coarse gravel, sand and silt with good gradation. These
deposits host the main aquifer system of the area. Recent Alluvial, the youngest
sediments exposed, occur along the Karaj and Kan streams, consists of gravel
and sand in northern plain and silt and clay in southern plain.

Geoelectrical Measurements and Interpretations

Vertical electrical soundings with a Schlumberger configuration in the Tehran-
Karaj plains have been conducted by [17]. 144 soundings in 11 profiles were
conducted in Shahriar plain. The apparent resistivity data obtained for different
values of AB/2 have been processed. These results were subsequently used to
obtain a realistic picture of the geological and hydrogeological framework.

Based on different particle size in unconsolidated horizons, electrical
resistivity varies from 10 - 250 Ωm. Lower resistivities occur in central and
southern parts of study area, where as clay and silt zones are predominant.
Data derived from lithology of boreholes and drilled wells located in the
vicinity of corresponding soundings were compared with geoelectrical surveys
results to obtain the true resistivity values, the thickness and the limitations
of the alluvial aquifer. The resultant isopach map (**Figure 5**(a)) shows that
aquifer thickness varies from a few meters to about 250 m. It increases from
25 - 100 m in northern edge of the plain to 100 - 250 m in the central parts of
plain. Aquifer thickness decreases toward south of area where it reaches to its
minimum.

Hydrogeologial Setting

The rapid increase in the population, caused by migration of people from adjacent provinces to the Tehran province, has led to large-scale groundwater developments in the study area over past two decades. As a result, aquifer condition has been changed egregiously during these years.

Based on time-drawdown and recovery data from 8 observation wells, transmissivity is about 750 $m^2 \cdot d^{-1}$ in northern parts and up to 3000 $m^2 \cdot d^{-1}$ in central parts of the plain (Figure 5(b)). It decreases again toward south where it is about 500 $m^2 \cdot d^{-1}$ in outlet of basin. Storage coefficient ranges between 0.25 - 0.35.

Drilled wells logs, water level in observing wells and geophysical investigations shows that main aquifer of Shahriar plain is an unconfined aquifer whose alluvial particle size varies from gravel and sand in north to silt and clay in south. The groundwater occurrence changes according to topographical and subsurface geological conditions. For example the water level is over 100 m below groundwater in the northernmost parts of the area and gradually gets closer to the ground surface toward the south of outlet of plain where it is about 10 m below ground level (Figure 5(c)).

Water levels of 30 observing wells have been used to draw isopotential maps of groundwater level in Shahriar plain. Based on this map, ground water flows from the northern heights and recharge areas of Karaj and Kan rivers toward the central parts and finally discharges to the outlet of the plain. The hydraulic gradient in the study area has its maximum value (13.5% - 18%) in north. While in central zone, due to increasing of saturated thickness of aquifer, its minimum values are found. Towards the study area's southern border and outlet of plain, hydraulic gradient increases again (3% - 8%).

Long-term and seasonal average areal water-level fluctuations recorded in 30 observing wells are plotted during the period 1989-2002. The mean values are calculated by creating Thiessen polygons for observation well stations. According to hydrograph analysis of observing wells it is concluded that there is an average decline of about 13.2 m in the groundwater level. Figure 4shows the water level fluctuation of the Shahriar plain.

According to Tehran Regional Water Authority [18], pumping wells are the main and only way to groundwater exploitation. All qanats and springs existing in the area have been dried in recent years due to intensive water level decline. The total withdrawal from deep and shallow pumping wells are 223, 57 and 38 million $m^3/$ year for agriculture, drinking and industrial pumping wells respectively.

Groundwater Balance

In this study, a groundwater balance has been prepared for the Shahriar aquifer based on available inflows, outflows and changes in the aquifer groundwater storage. The study was conducted for the period 1999 - 2000. In a specified period changes in volume storage of an aquifer depends on average flows entering and leaving the system through the different sources and sinks. Therefore the groundwater balance has two components: the total

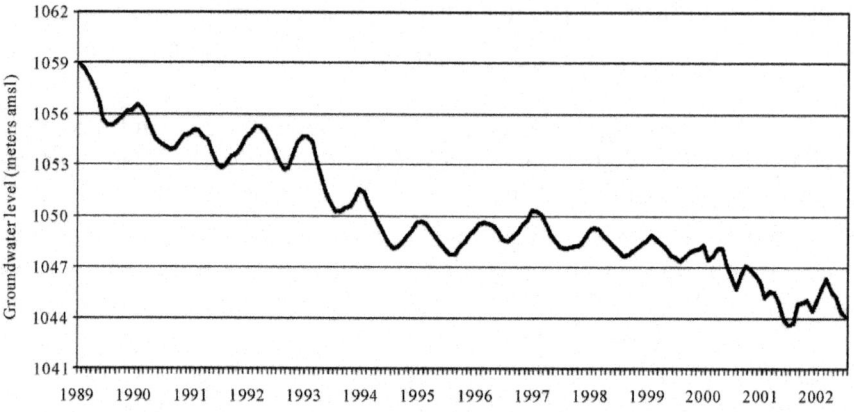

Figure 4: Groundwater level fuctuation of the Shahriar aquifer (Data derived from 30 monitoring boreholes).

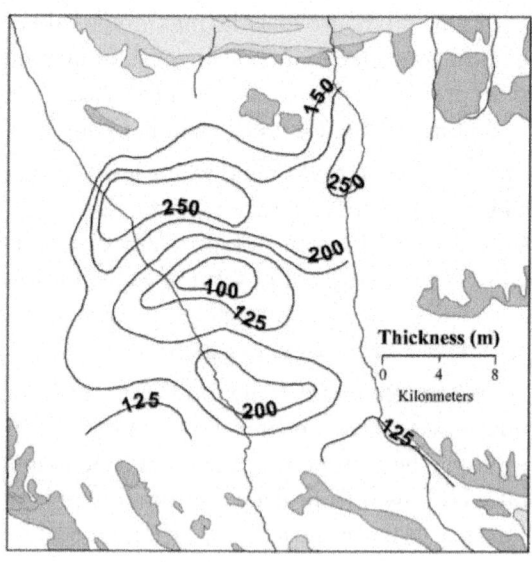

(a)

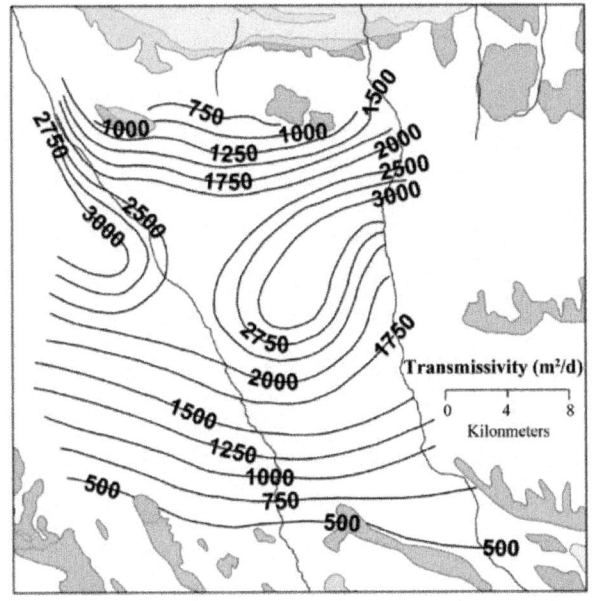

(b)

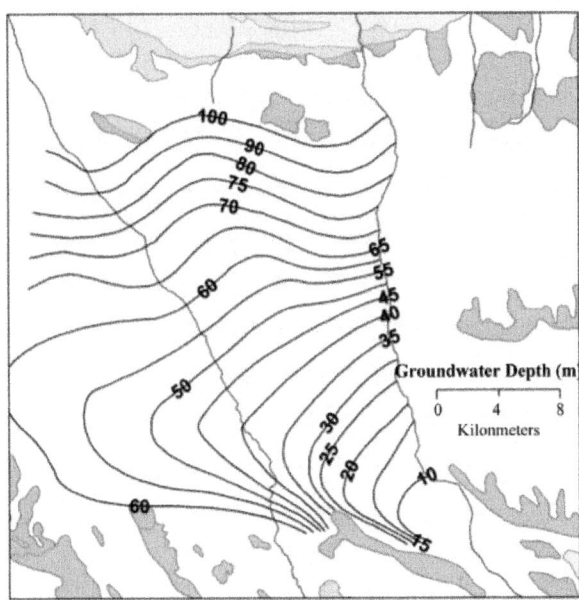

(c)

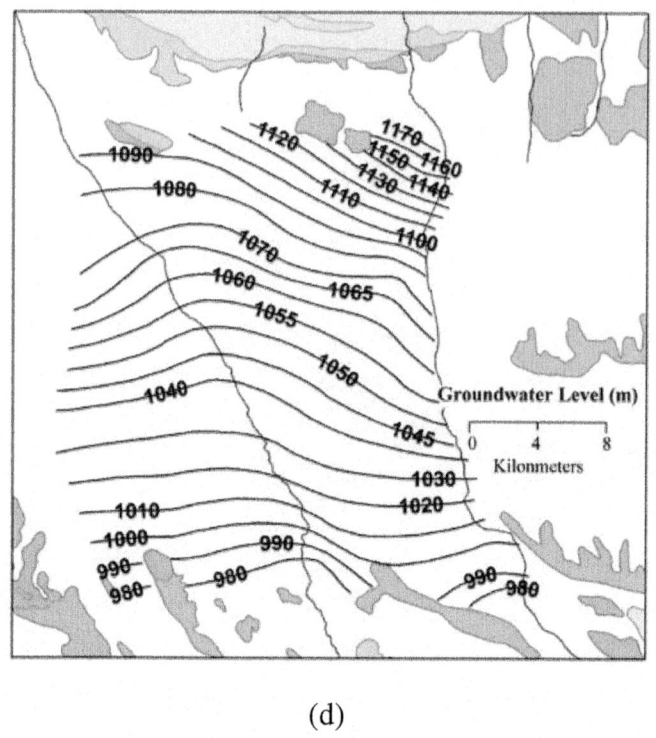

(d)

Figure 5: Hydrogeological maps of the study area: (a) Thichness isopack map; (b) Transmissivity; (c) Groundwater table depth; (d) Groundwater level.

input (ΣQ_{in}) and the total output (ΣQ_{out}). The equation for the groundwater balance during a specified hydrological period can be written as [19]

$$\Sigma Q_{in} - \Sigma Q_{out} = \pm \Delta V \tag{1}$$

where ΔV is the change in groundwater storage.

In Shahriar plain the total groundwater input (ΣQ_{in}) consists of: lateral subsurface inflows (Q_1), rainfall recharge (Q_2), Mountain-front recharge (Q_3), recharge due to irrigation returns (Q_4), recharge due to domestic and industrial wells returns (Q_5) and streambeds and canal infiltrations (Q_6)

$$\Sigma Q_{in} = Q_1 + Q_2 + Q_3 + Q_4 + Q_5 + Q_6 \tag{2}$$

The total groundwater output (ΣQ_{out}) consists of subsurface outflows (Q_7) and pumping groundwater (Q_8)

$$\Sigma Q_{out} = Q_7 + Q_8 \tag{3}$$

The change in groundwater storage (ΔV) can be estimated as

$$\Delta V = A \cdot S_c \cdot \Delta h \tag{4}$$

where:

A = Area of aquifer

Δh = Difference between the average water level at during the hydrological period and S_c = Average storativity of aquifer

RESULTS AND DISCUSSION

Groundwater recharge and discharge in the Shahriar aquifer occurs via seven and two components respectively. These components are contributing in groundwater system as explained below:

Lateral Subsurface Inflows (Q_1)

Inflow and outflow cross sections were determined using equipotential groundwater level map (**Figure 5**(d)) for October 1999 (23 cross sections). Employing Darcy's law and transmissivity map, an estimate of the volume of lateral inflows from the upstream boundary of the aquifer was performed:

Inflow amount (Q_1) = average transmissivity (T) × cross sectional length (L) × gradient (i) × water balance period (t).(5)

For each cross section volume of inflow was calculated using mentioned equation. The total lateral inflows to the aquifer system were calculated to be $Q_1 = 167$ Million $m^3 \cdot yr^1$ for the hydrological year 1999-2000.

Rainfall Recharge (Q_2)

Various parameters such as annual rainfall depth, slope and soil permeability affect the rainfall recharge amount. Accepting annual rainfall depth about 225 mm for year 1999-2000 and the area about 580 km^2 the annual rainfall volume equals 130 million m^3.

The infiltration ratio considering the lithology and soil texture is estimated 10% of the annual rainfall. Thus in this study the annual recharge of the aquifer system from rainfall was estimated to be $Q_2 = 13$ million $m^3 \cdot yr^{-1}$.

Mountain-Front Recharge (Q_3)

As mentioned in hydrology section, the annual volume of entering runoff to study area via three sub-basins named Kan-Chitgar, Chitgar, and Chitgar-Karaj are about 1.5, 13.5 and 3 respectively and totally about 18 million m^3. The total recharge through these runoffs is estimated about 15 million m^3.

Recharge Due to Irrigation Returns (Q_4)

According to FAO technical paper about irrigation yield and percolate water losses in farms based on irrigation method and soil texture [20], the irrigation returns are estimated to be as high as 25% of the applied irrigation (223 million $m^3 \cdot yr^{-1}$ from agricultural wells, 29 million $m^3 \cdot yr^{-1}$ through water pumping from streams and canals). Hence, recharge from irrigation returns is 63 million $m^3 \cdot yr^{-1}$.

Recharge Due to Domestic and Industrial Wells Returns (Q_5)

Conciderably percent of water applying for drinking and household uses, turns to wastewater and to some extent percolate to the aquifer. Groundwater extracted for municipal uses from the region aquifer is about 57 million $m^3 \cdot yr^{-1}$. The amount of return flow from these well is estimated about 37 million $m^3 \cdot yr^{-1}$ (65% of the total amount).

The total amount of groundwater abstracted for industrial uses is about 38 million $m^3 \cdot yr^{-1}$ which is estimated 35% of this amount (13 million $m^3 \cdot yr^{-1}$) percolate to the aquifer. Therefore the recharge due to domestic and industrial wells calculated as $Q_5 = 37 + 13 = 50$ million $m^3 \cdot yr^{-1}$.

Streambed and Canals Infiltrations (Q_6)

The total volume of inflow to study area through Navvab canal is 24 million $m^3 \cdot yr^{-1}$. From this amount 5 million $m^3 \cdot yr^{-1}$ is being used for irrigation that 1.3 million $m^3 \cdot yr^{-1}$ (25%) participates in aquifer recharging. From the rest of total volume (19 million $m^3 \cdot yr^{-1}$), 1.9 million $m^3 \cdot yr^{-1}$ (tantamount 10%) infiltrates to aquifer system.

Eight Streams are taking supply from Bileghan diversion dam in north west of study area. The recharge to the aquifer system from streambed infiltration was estimated to be 2.1 million $m^3 \cdot yr^{-1}$ (10% of total streams flow). From the estimates made above, the annual input to the Shahriar alluvial aquifer is $\Sigma Q_{in} = Q_1 + Q_2 + Q_3 + Q_4 + Q_5 + Q_6 = 328.3$ million $m^3 \cdot yr^{-1}$

The output parameters of the aquifer are groundwater extraction and groundwater discharge across sub-basin boundaries. According to water table depth map of study area which shows minimum depth of water table is more than 5 meter in the region, Evaporation from groundwater has been considered negligible.

Lateral Subsurface Outflows (Q_7)

The total lateral subsurface outflows to the aquifer system were calculated to be $Q_7 = 20$ million $m^3 \cdot yr^{-1}$ according to Darcy's law as it mentioned in lateral inflows.

Pumping Groundwater (Q_8)

Groundwater exploitations are performed via numerous production wells in order to cover the regional needs as below.

Based on the data of pumping wells from the Tehran

Table 1: Groundwater balance

Water input (million $m^3 \cdot yr^{-1}$)		Water output (million $m^3 \cdot yr^{-1}$)	
Lateral groundwater fluxes	167	Subsurface outflows	20
Recharge due to rainfall	13	Agricultural abstractions	223
Mountain-front recharge	15	Domestic use	57
Irrigation return	63	Industrial use	38
Domestic and industrial wells returns	50		
Streambed and canal infiltration	5.3		
	313.3		338

Regional Water Board inventory report, groundwater abstracted from the Shahriar aquifer is 223, 57 and 38 million $m^3 \cdot yr^{-1}$ for agricultural, Domestic and industrial consumptions respectively.

From the estimates made above, the mean annual output from the Shahriar alluvial aquifer is $\Sigma Q_{out} = Q_7 + Q_8 = 20 + 318 = 338$ million $m^3 \cdot yr^{-1}$

As illustrated in Table 1, a negative groundwater balance has been calculated in the studied aquifer system for the year 1999-2000. Groundwater overexploitation by abstractions from over 3500 production wells has resulted in considerable head decline that reached 1.42 m during the hydrological year 1999-2000. It can be concluded that the current abstraction is not sustainable, thus groundwater is withdrawn from storage. The impacts of the overexploitation are the drying up of upper aquifer system wells and qanats, the land subsidence [21,22] and the severe deterioration of groundwater quality.

There are some strategies in order to make the Shahriar groundwater system sustainable. Groundwater exploitation controlling based on plant water demand and restriction the irregular pumping wells' drilling can play an important role in decreasing the depletion of water table. Utilization of the treated wastewater and the application of water-saving techniques such as spray irrigation and drip irrigation can decrease the groundwater quantities for irrigation use [23,24].

Artificial recharge is another alternative as a valuable water management tools that effectively help to offset increased demands for water [25]. Availability, quality and quantity of source water available, resulting water quality (reactions with native water and aquifer materials), clogging potential, underground storage space available, depth to underground storage space, transmission characteristics, and costs are some factors controlling the feasibility of artificial recharge method [26]. The lithological studies in the study area reveals that the northern parts especially in Chitgar and the region between Saeid Abad, Ghale Hasan Khan and Shariar can be considered as suitable recharging zones due to thick and permeable horizons of coarse grain deposits.

Therefore, an appropriate set of management strategies, including water conservation measures, regulation of existing development, improvement of current legislation and public education should be adopted [3]. Future investigations of the groundwater recharge in the alluvial aquifer system would benefit by improvement in hydrological data monitoring, the application of isotopic analysis and computer modeling to simulate the water cycle and groundwater flow.

ACKNOWLEDGEMENTS

The authors wish to thank Tehran Regional Water Authority for supplying the existing relevant data. Also, the constructive comments of anonymous reviewers will be appreciated.

REFERENCES

1.   E. Wendland, C. Barreto and L. H. Gomes, "Water Balance in the Guarani Aquifer Outcrop Zone Based on Hydrogeologic Monitoring," Journal of Hydrology, Vol. 342, No. 3-4, 2007, pp. 261-269. doi:10.1016/j.jhydrol.2007.05.033

2.   J. C. Marechal, B. Dewandel, S. Ahmed, L. Galeazzi and F. K. Zaidi, "Combined Estimation of Specific Yield and Natural Recharge in a Semi-Arid Groundwater Basin with Irrigated Agriculture," Journal of Hydrology, Vol. 329, No. 1-2, 2006, pp. 281-293.doi:10.1016/j.jhydrol.2006.02.022

3.   A. T. Tizro, K. S. Voudouris and M. Eini, "Groundwater Balance, Safe Yield and Recharge Feasibility in a Semi- -Arid Environment: A Case

Study from Western Part of Iran," Journal of Applied Sciences, Vol. 7, No. 20, 2005, pp. 2967-2976.

4.   G. R. Lashkaripour, "Contamination of Groundwater Resource in Zahedan City Due to Rapid Development," Journal of Applied Science, Vol. 3, No. 5, 2003, pp. 341-345.doi:10.3923/jas.2003.341.345

5.   G. B. Allison, "A Review of Some of the Physical Chemical and Isotopic Techniques Available for Estimating Ground Water Recharge," In: I. Simmers, Ed., Estimation of Natural Ground Water Recharge, Reidel, Dordrecht, 1988, pp. 49-72.

6.   S. S. D. Foster, "Quantification of Ground Water Recharge in Arid Regions: A Practical View for Resource Development and Management," In: I. Simmers, Ed., Estimation of Natural Ground Water Recharge, Reidel, Dordrecht, 1988, pp. 323-338.

7.   R. W. Healy and P. G. Cook, "Using Groundwater Levels to Estimate Recharge," Hydrogeology Journal, Vol. 10, No. 1, 2002, pp. 91-109. doi:10.1007/s10040-001-0178-0

8.   C. Leduc, J. Bromley and P. Schroeter, "Water Table Fluctuation and Recharge in Semi-Arid Climate: Some Results of the HAPEX-Sahel Hydrodynamic Survey (Niger)," Journal of Hydrology, Vol. 188-189, 1997, pp. 123-138. doi:10.1016/S0022-1694(96)03156-3

9.   Tehran Regional Water Authority Website, 2009. http://www.tw.org.ir/dams/selectdam_en.asp

10. Tehran Regional Water Authority, "Annual Report of Groundwater Resources Exploitation," Deputy for Water Resources Study and Research, Tehran Regional Water Authority, 2005.

11. Water Resources Studies Division, "Climatology and Hydrologic Report of Tehran, Varamin and Shahriar Basins," Tehran Regional Water Authority, 1997.

12. A. Alizadeh, "The Principle of Applied Hydrology," Astan Ghods Razavi Publications, Mashhad, 2009.

13. A. Darvishzadeh, "Geology of Iran," Amirkabir Publication, Tehran, 2003.

14. S. A. Aghanabati, "Geology of Iran," Geological Survey of Iran Publication, Tehran, 2004.

15. C. Vita-Finzi, "Late Quaternary Alluvial Chronology of Iran," Geologische Rundschau, Vol. 58, No. 2, 1969, pp. 951-973. doi:10.1007/BF01820740

16. E. H. Rieben, "The Geology of the Tehran Plain," American Journal of Science, Vol. 253, 1955, pp. 617-639. doi:10.2475/ajs.253.11.617

17. Water and Soil Engineering Unit of Jihad-e-Sazandegi, "Geoelectrical Studies in Tehran-Karaj Plain," Jihad-eSazandegi Ministry, Tehran, 1983.

18. Tehran Regional Water Authority, "Detailed Data Collection from Discharges of Pumping Wells and Qanats in Tehran Province," Tehran Regional Water Authority, Tehran, 2001.

19. J. J. Carrillo-Rivera, "Application of the Groundwater-Balance Equation to Indicate Interbasin and Vertical Flow in Two Semi-Arid Drainage Basins, Mexico," Hydrogeology Journal, Vol. 8, No. 5, 2000, pp. 503-520. doi:10.1007/s100400000093

20. A. Alizadeh, "Relation between Water, Soil and Plant," Astan Ghods Razavi Publications, Mashhad, 2007.

21. A. Shemshaki and I. E. Soltani, "Land Subsidence Review at Sahariar Plain," Geological Survay of Iran, Tehran, 2004.

22. R. Umar, M. M. A. Khan, I. Ahmed and Sh. Ahmed, "Implications of Kali–Hindon Inter-Stream Aquifer Water Balance for Groundwater Management in Western Uttar Pradesh," Journal of Earth System Science, Vol. 117, No. 1, 2008, pp. 69-78. doi:10.1007/s12040-008-0014-1

23. P. K. Naik, and A. K. Awasthi, "Groundwater Resources Assessment of the Koyna River Basin, India," Hydrogeology Journal, Vol. 11, No. 5, 2003, pp. 582-594.doi:10.1007/s10040-003-0273-5

24. M. N. Bhutta, M. Saeed and M. Rafiq, "Evaluation of Groundwater Balance—A Case Study of Mona Drainage Basin," Pakistan Journal of Water Resources, Vol. 11, No. 2, 2007, pp. 19-26.

25. K. Voudouris, P. Diamantopoulou, G. Giannatos and P. Zannis, "Groundwater Recharge via Deep Boreholes in the Patras Industrial Area Aquifer System (NW Peloponnesus, Greece)," Bulletin of Engineering Geology and the Environment, Vol. 65, No. 3, 2006, pp. 297-308. doi:10.1007/s10064-005-0036-8

26. M. P. O'Hare, D. M. Fairchild, P. A. Hajali and L. W. Canter, "Artificial Recharge of Groundwater: Status and Potential in the Contiguous United States," Lewis Publishers, Chelsea, 1986.

Chapter 9

CHANGING HYDROLOGY OF THE HIMALA-YAN WATERSHED

Arshad Ashraf[1]

[1]Water Resources Research Institute, National Agricultural Research Center, Islamabad, Pakistan

INTRODUCTION

The Himalayan region is a source of ten major river systems that together provide irrigation, power and drinking water for 1.3 billion people i.e. over 20% of the world's population. The supply and quality of water in this region is under extreme threat, both from the effects of human activity and from natural processes and variation [1]. Population growth is already putting massive pressure on regional water resources, affecting water resource in terms of demand, water-use patterns and management practices. The change in hydrological cycle may affect river flows, agriculture, forests, biodiversity and health besides creating water related hazards [2]. The need for suitable strategies for climate resilient development has policy and governance implications [3]. Adaptation to climate change is the area that should be strengthened through policy advocacy supported by evidence through rigorous research and verified information.

Re-assessment of true catchments yields under existing and future scenarios of landuse and climate changes is very essential to devise watershed management strategies which can minimize adverse impacts both in terms of quantity and quality. Since trends are still unclear, the extent to which changes can be attributed to variable environmental changes is difficult to determine. It has become imperative to assess ongoing hydrological changes and changes that might occur in future to devise appropriate adaptation measures to foster resilience to future climate change, thereby enhancing water security.

In the present study, SWAT model developed by United States Department of Agriculture (USDA) [4] has been used to evaluate surface runoff generation, soil erosion and quantify the water balance of a Himalayan watershed in the

Northern Pakistan. The response of watershed yield to historical landuse evolution and under variable landuse and climate change scenarios has been studied in order to mitigate the negative impacts of these changes and promote development activities in this region. The study would provide basis to recommend changes in the water management regimes so as to address future adaptation issues.

Modeling Hydrological Processes

Dealing with water management issues requires analyzing of different elements of hydrologic processes taking place in the area of interest. As such processes are taking place in a combine system that exists at a watershed level, thus the analysis must be carried out on a watershed basis. Understanding of relationship between various watershed characteristics such as morphology, landuse and soil, and hydrological components are very essential for water resources development in any area. Since the hydrologic processes are very complex, their proper comprehension is essential and for this watershed models are widely used. Most of the watershed models basically simulate the transformation of precipitation into runoff, sediment outflow and nutrient losses. Changes in landuse including urbanization and de(/re)forestation continue to affect the nature and magnitude of surface and subsurface water interactions and water availability influencing ecosystems and their services. One can formulate water conservation strategies only after understanding the spatial and temporal variations and the interaction of these hydrologic components. The alarming rate of soil erosion in context of changing landuse and climate in the Himalayan region calls for urgent attention for this problem. Assessment of erosion is a very difficult task when executed using conventional methods and requires to be done repetitively. The use of an appropriate watershed model is thus essential to deal with such problems.

Choice of watershed development model depends upon the hydrologic components to be incorporated in the water balance. The most important hydrologic elements from the water management point of view are surface runoff, lateral flow, baseflow and evapotranspiration. In presenting an appropriate view of reality, model must remain simple enough to understand and use. There are a number of integrated physically based distributed models, among which researchers have identified Soil and Water Assessment Tool (SWAT) as the most promising and computationally efficient [4]. The model is an integrated physically based distributed watershed model that has an ability to predict the impact of land management practices on water, sediment yield and agricultural chemical yield [5]. Distributed models also take the spatial variability of watershed properties into account.

Model Description

The SWAT is a process-based continuous daily time-step model that offers distributed parameter, continuous time simulation, and flexible watershed configuration [6]. It has gained international acceptance as a robust interdisciplinary watershed modeling tool. Two methods are used for surface runoff estimation in SWAT i.e. the SCS curve number and Green-Ampt infiltration. This study is based on the use of curve number for surface runoff and hence stream flow simulation. A SWAT model can be built using the Arc-View interface called AVSWAT which provides suitable means to enter data into the SWAT code. Main processes include water balance calculations (i.e. surface runoff, return flow, percolation, evapotranspiration, and transmission losses), estimation of sediment yield, nutrient cycling and pesticide movement.

The spatial heterogeneity is represented by means of observable physical characteristics of the basin such as landuse, soils and topography etc. Model inputs include physical characteristics of the watershed and its sub-basins i.e. precipitation, temperature, soil type, land slope, Manning's n values, USLE K factor, and management inputs like crop rotations, planting and harvesting dates, tillage operations, irrigation, fertilizer use, and pesticide application rates. Model outputs include sub-basin and watershed values for surface flow, ground water and lateral flow, sediment, nutrient and pesticide yields. The main basin is divided into sub-basins which are further divided into hydrologic response units (HRU) composed of homogeneous landuse, soil types, relevant hydrological components and management practices. Sediment yield is estimated by the Modified Universal Soil Loss Equation (MUSLE; [7]. The model has been applied worldwide for the purpose of simulating sediment flow [8], modeling hydrologic balance [9], evaluation of the impact of landuse and landcover changes on the hydrology of catchments [10]. The model provides a flexible capability for creating climate change scenarios evaluating a wide range of "what if" questions about how weather and climate could affect our systems.

Equations of Watershed Hydrology

The hydrologic process in a watershed is simulated by the following water balance equation:

$$SW_t = SW + \sum_{i=1}^{t} \left(R_i - Q_i - ET_i - P_i - QR_i \right)$$

(1)

where: SW_t is the final soil water content (mm), SW is the initial soil water content minus the permanent wilting point water content (mm), t is time in days, R is rainfall (mm), Qi is surface runoff (mm), ETi is evapotranspiration (mm), Pi is percolation (mm) and QRi is lateral flow (mm). The surface runoff is predicted by the following equation:

$$Q = \frac{(R-0.2s)^2}{R+0.8s} \text{ for } R > 0.2s$$

$$Q = 0.0 \quad \text{for } R < 0.2s$$

(2)

$$s = 254\left(\frac{100}{CN} - 1\right)$$

(3)

Where, Q = daily surface runoff (mm): R = daily rainfall (mm), S = retention parameter (mm); CN = curve number.

Lateral flow is predicted by:

$$q_{lat} = 0.024\frac{(2SSC\sin\alpha)}{\theta_d L}$$

(4)

Where, qlat = lateral flow (mm/ day); S = drainable volume of soil water per unit area of saturated thickness (mm/day), SC = saturated hydraulic conductivity (mm/h); L = flow length (m); α = slope of the land: ɵd = drainable porosity

The base flow is estimated by:

$$Q_{gwj} = Q_{qwj-1} \cdot e^{(-\alpha_{gw} \cdot \Delta t)} + w_{rchrg} \cdot \left(1 - e^{(-\alpha_{gw} \cdot \Delta t)}\right)$$

(5)

Where, Qgwj = groundwater flow into the main channel on day j; αgw = base flow recession constant; Δt= time step. The computed runoff from each element is integrated using a finite difference form of the continuity equation relating moisture supply, storage and outflow.

Description of Study Area

Rawal watershed covers an area of 272 sq km within longitudes 73° 03′ - 73° 24′ E and latitudes 33° 41′ - 33° 54′ N comprising parts of Margalla hills and Murree mountains in the southern Himalayas of Pakistan (Figure 1). About 47% of the watershed area lies in the Islamabad Capital Territory while the rest in Punjab and Khyber Pakhtunkhwa (KPK) provinces, so it is well connected through a metalled road with other parts of the country. Korang is the main river flowing in the watershed that receives runoff from watershed via four major and 43 small streams [11]. Rawal dam is constructed on Korang river, which supplies 22 million gallons per day of water for drinking and other household needs to Rawalpindi city and a limited water for irrigation use to Islamabad area. The elevation ranges between 523 meters and 2145 meters above mean sea level (masl). Physiographically, the watershed comprises of 34% hilly area (Elev. <700 masl), 62% Middle mountains (Elev. within 700-2000 masl) and 4% High mountains (Elev. >2000 masl).

The Himalayas serve as a divide between Central Asia and South Asia. The Indo-Eurasian plates collision resulted in the formation of new relief and topography, which consists of series of mountain ranges located in the north and west of Pakistan, commonly known as the Himalayan Mountain System [12]. The principal uplift occurred during the middle or late Tertiary period, 12 to 65 million years ago. The study area lies in sub-humid to humid sub-tropical continental highlands. The hottest months are May, June and July. The mean maximum temperature ranges between 17.6°C and 40.1°C while mean minimum temperature between 2.1°C and 21.6 °C. The winter months are from October to March. The highest temperature was recorded as 46.6°C in 2005 and the lowest as -3.9°C during 1967 [13]. Mean annual rainfall of 1991-2010 period is about 1232 mm. The occurrence of rainfall is highly erratic both in space and time. Over 60 percent of the annual rainfall occurs during monsoon season i.e. from July to September. Most of the rainfall is drained out rapidly due to steep slopes and dissected nature of the terrain. Springs and streams are the main source of water for drinking and other domestic requirements. A prolonged dry season may cause water shortage in some parts of the area.

Underlying rocks consist of poorly compressed and highly folded and faulted Murree series that are moderately to severely eroded, shallow clayey loams of very low productivity [14]. The soils formed over shale are clayey while those developed on the sandstone are sandy loams to sandy clay loam in texture. The flora is mainly natural with xeric, broad-leaved deciduous, evergreen trees and diverse shrubs on the southern slopes. The dominating plant species are Carissa spinarum (Granda), Dodonaea viscosa (Sanatha) and Olea ferruginea (Wild Olive). Sub-tropical pine zone occupies steep and very steep mountain slopes [15]. Agriculture is practiced in small patches of land as terrace cultivation.

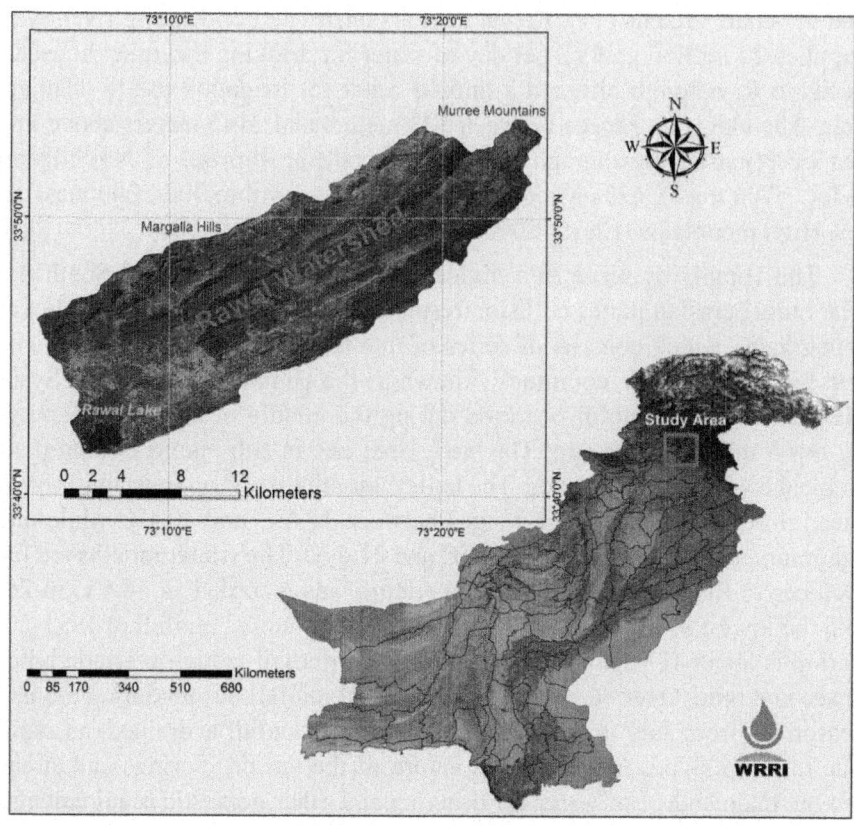

Figure 1: Location map of the study area.

Main Environmental Issues

The watershed is confronting problems of rapid urban development and deforestation due to which its landuse is changing gradually. The population growth and addition of a number of housing colonies in the Rawal Lake catchment area are adversely affecting the regime of water coming into Rawal Lake. The activities like illegal cuttings due to high market value of forest wood and intensive use of forest wood for household needs (cooking, heating, timber etc.), ineffective forest management and forest disease etc. are accelerating the deforestation rate in the watershed area [16-17]. Destruction of aquatic habitat and a reduction of water quality are some of the negative impacts of deforestation. Extensive cattle grazing and fuel wood cutting by the local communities have deformed the plants to bushes [18]. The removal of a forest cover from steep slopes often leads to accelerated surface erosion and dramatically increases the frequency of land sliding and surface runoff.

The storage capacity of the Rawal Lake which was 47,230 acre-ft when it was developed in 1960, has been reduced almost 34 percent due to sedimentation generating from natural and human induced factors in its catchment area [16]. The use of pesticides and herbicides in agriculture is a source of toxic pollution [19]. Many housing schemes, recreational pursuits e.g. Lake view point, Chatter and Valley parks etc. and farmhouses have been developed in the watershed. The construction of roads, pavements and other structures reduce the infiltration area that ultimately affect the recharging of the aquifer of the twin cities. No systematic study has been undertaken yet to document the landuse variability in the watershed.

MATERIALS AND METHODS

Data Used

In the present study, the basic watershed data used to extract spatial input for SWAT model were hydrologic features, soil distribution, landuse information, and topography. The remote sensing technique has potential application in landuse monitoring and assessment at desired scales. RS images of LANDSAT-TM (Thematic Mapper) of period 1992 and LANDSAT-ETM+ (Enhanced Thematic Mapper Plus) of 2000 and 2010 periods (Path-Row: 150-37) were used to delineate landuse/landcover of the watershed area on temporal basis. The LANDSAT ETM+ sensor is a nadir-viewing, 7-band plus multi-spectral scanning radiometer (upgraded ver. of TM sensor) that detects spectrally filtered radiation from several portions of the electromagnetic spectrum. The spatial resolution (pixel sizes) of the image data includes 30 m each for the six visible, near-infrared, and short-wave infrared bands, 60 m for the thermal infrared band, and 15 m for the panchromatic band. The climatic parameters i.e. daily temperature (max& min) and precipitation data recorded at Satrameel observatory (73° 12′ 50" E, 33° 45′ 57" N & Elev: 610 m) maintained by Water Resources Research Institute had been collected for period 1991-2010. The discharge data of Korang river available on monthly basis from Small dams organization was acquired for the same period for model calibration and validation. The soil map developed by Soil Survey of Pakistan was utilized to extract soil data attributes for the study area.

Data Preparation

The base map of the study area was prepared through generating and integrating thematic layers of elevation, physiography and infrastructure using ArcGIS 9.3 software. An integrated hydrological, spatial modeling and field investigations approach was adopted to achieve the study output. The boundaries of the watershed and sub-basins were delineated using Aster 30m DEM of the area in SWAT model 2005 software. Elevation map comprising of four classes i.e. >1600m, 1200-1600m, 800-1200m and <800m range, was prepared from Aster 30m DEM data (Figure 2). The image data was georeferenced using Universal Transverse Mercator (UTM) coordinate system (Zone 43 North). The satellite images were analyzed through visual and digital interpretation techniques to observe spatial variability of landuse. The visual interpretation was performed for qualitative analysis while digital interpretation for quantitative analysis of the image data. The false color composite of 5, 4, 2 (RGB) of LANDSAT image data was selected to extract signatures of representative landcover types from the image. In this bands combination, landcover is visible in true color i.e. vegetation in green, soil in pale to reddish brown and water in shades of blue color. The built-up area is shown in mixed pattern of white, brown, and purple colors due to variable types and density of constructed area, mixing of new and old settlements, presence of land features like lawns, parking sides, water ponds, roads/tracks etc. The signatures were evaluated using error matrix and an overall accuracy of more than 95 percent was achieved.

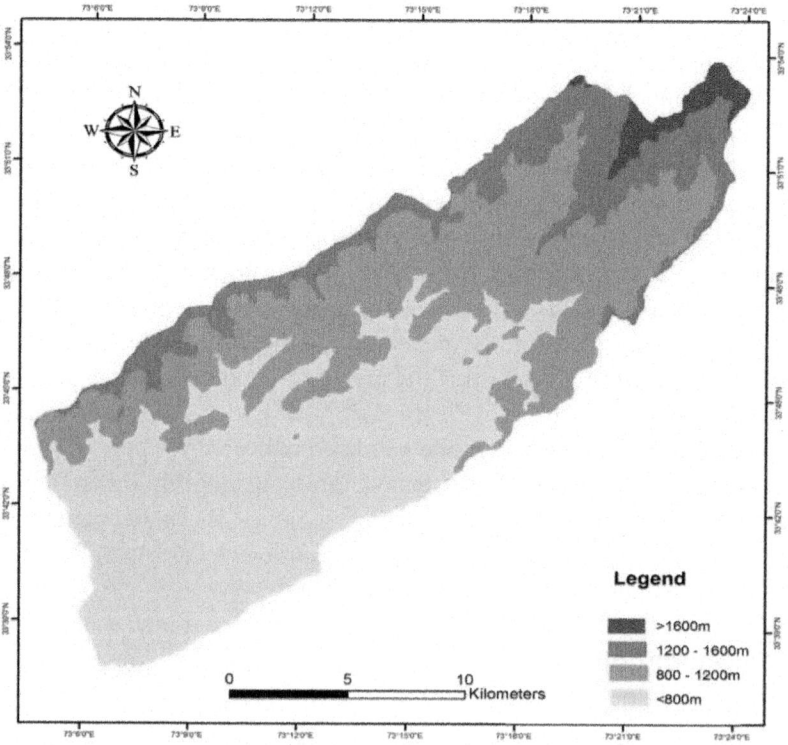

Figure 2: Elevation increases gradually towards northeast in the study area.

The classification of the images was performed using supervised method following maximum likelihood rule mostly used to acquire reliable classification results. The classification output was supported with Normalized difference vegetation index - NDVI data that helps in segregating vegetative areas from non-vegetative [20]. The index which is based on the spectral characteristics of green vegetation cover in the area uses TM3 and TM4 bands of LANDSAT ETM+ image as given in the following equation:

$$NDVI = (TM4 - TM3) / (TM4 + TM3) \tag{6}$$

The classification of the images was performed to obtain seven major landuse/landcover classes which include conifer forest, scrub forest, agriculture, rangeland, soil/rocks, settlement and water. The images were recoded and later filtering technique was applied to remove noisy/misclassified pixels from the recoded image data. The doubtful classes were modified after ground truthing i.e. performing field survey in the target areas. Finally change analysis of landuse/landcover classes was performed using spatial modeling functions of ERDAS Imagine 9.2 software.

Model Baseline Establishment

Main procedures in the model running includes: (a) development of streams and sub-basins databases, (b) landuse and soil data input within sub-basins, (c) Input variable parameters of climate and management options, (d) compilation of input data and running the model for generating output results. The entire watershed had been divided into 15 sub-basins by choosing a threshold area of 500 ha. A total of 73 HRUs were generated in those sub-basins. A threshold of 5% was defined landuse distribution and 15% for soil distribution over sub-basin area. The low percentage for landuse was used to accommodate conifer coverage distributed in patches over northeastern parts of the watershed area. The importance of land uses lies mainly in the computation of surface runoff with the help of SCS curve during the model operation [6]. Three soil classes were identified and mapped i.e. sandy clay loam over northwestern hilly terrain, sandy clay loam over valley area in the northeast and sandy loam over low plains in the southwestern part of the watershed area.

The subcomponents of the water balance identified for use in analyses are total flow (water yield) consisting of surface runoff, lateral and base flow, soil water recharge; and actual evapotranspiration. These components are expressed in terms of average annual depth of water in millimeters over the total watershed area. For estimation of sediment yield, C factor values were used on the basis of soil erosion study [21] carried out previously in Pothwar region. The C value of 0.176 was assigned to soil/rocks while 0.2 was assigned to agricultural land class. Higher C values indicate more risk of soil erosion. The conservation practice factor P was assigned value of 1 on account of no significant conservation practice present in the watershed area [22].

Model Calibration

Calibration and validation of the SWAT model was performed using monthly river flows data of 1991-2010 period. Data pertaining to year 1991 to 2006 had been used for calibration and the rest for validation of the model. The purpose of model validation is to assess whether the model is able to predict field observations for time periods different from the calibration period [23]. Although the model was run for years 1991 to 2006, the first 6 years of the simulated output were disregarded in the calibration process, since these are required by the model as a warm-up period. This period is essential for the stabilization of parameters (e.g groundwater depth), as the results sometimes vary significantly from the observed values. Thus the final calibration period was from January 1997 to December 2006. The calibration accuracy was checked by calculating several indexes which include Nash & Sutcliffe coefficient (NTD), Root Mean Square Error (RMSE) and the correlation coefficient R^2 of the time series. The Nash & Sutcliffe coefficient [24] is an estimate of the variation of a time series from another as given by following equation:

$$NTD = 1 - \frac{\sum_{i=1}^{n}\left(Q_{obs,i} - Q_{sim,i}\right)^2}{\sum_{i=1}^{n}\left(Q_{obs,i} - \overline{Q}_{sim,i}\right)^2}$$

(7)

And root mean square error- RMS was computed using following equation:

$$RMSE = \sqrt{\frac{1}{n}\sum_{i=1}^{n}\left[Wi\left(Q_{sim,i} - Q_{obs,i}\right)\right]^2}$$

(8)

Where, Q_{sim} = simulated time series, Q_{obs} = observed time series, Q_{sim} = numerical mean for the simulated time series, W = weight and n = total number of measurements. A Nash & Sutcliffe coefficient approaching unity indicates that the estimated time series is almost identical to the observed one. The results of these tests are summarized in Table 1. The NTD index reached the value of 0.80, signifying a quite precise calibration. Later the model was validated using the same indexes, for the period of January 2007 to December 2010. The results of statistical analysis indicated a Nash Sutcliff efficiency of 0.80. The simulated river flows matched well with the observed values (Figure 3).

TABLE 1: Criteria for examining the accuracy of calibration and validation processes.

Index	Calibrated period	Validated period
NTD	0.80	0.80
RMSE (mm)	17.0	30.4
$R2$	0.81	0.91

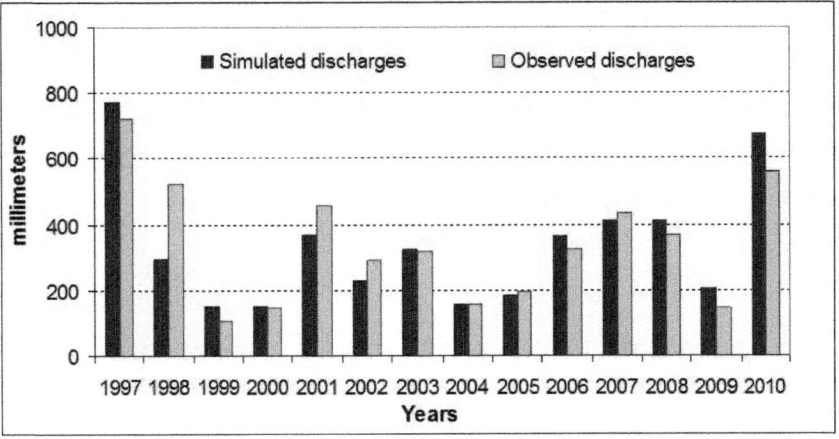

Figure 3: Time series of simulated and observed annual discharges for the Rawal watershed, period 1997-2010.

RESULTS AND DISCUSSION

Assessment of Changing Landuse/Landcover

Comprehensive information on the spatial and temporal distribution of landuse/landcover is essential for estimating hydrological changes at watershed level. The landuse/landcover condition of the watershed was estimated for three different periods i.e. 1992, 2000 and 2010 (Figure 4). Major landcover change was observed in the scrub class which indicated a reduction of about 4,515 ha during 1992-2010 period (Table 2). The rate of decrease in its coverage was about 1.5% per annum. The scrub wood is mostly used as fuel at local level due to non-availability of other energy sources. Major part of it had been converted into agriculture and built-up land, while in some areas it has changed into rangeland due to extensive wood cutting. These results are verified by the findings of [25] which highlighted maximum decrease in scrub forest during 30-year period i.e. 1977-2006 in Rawalpindi area. The settlement class had shown almost four times increase in coverage i.e. from 2.6% in 1992 to 8.7%

in year 2010. The average rate of increase in this class was about 90 ha y-¹. The rate was over 45 ha y-¹ during 1992-2000 while it was about 125 ha y-¹ during 2000-2010 period indicating rapid urbanization in the last decade (Figure 5). The conifer forest had shown a decline at a rate of about 2.1% y-¹ within last two decades. Although FAO [26] had reported deforestation at a rate of about 1.5% annually in the country, but due to high urban development, the rate of forest decline was higher in the watershed area. The agriculture coverage indicated an average increase of about 26 ha annually during 1992-2010 period. The rate of increase was about 3.4% y-¹ during 1992-2000 while it was 0.3% y-¹ during 2000-2010 period. The situation indicates intense agriculture activity in the former decade that seems replaced by rapid growth in urban development in the later decade.

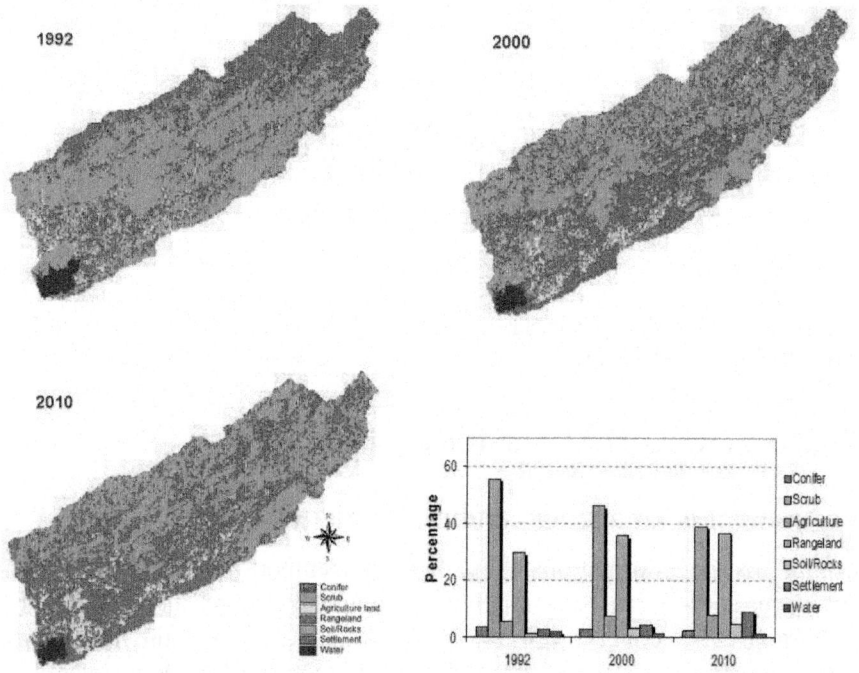

Figure 4: Spatio-temporal variations in landuse/landcover in Rawal watershed area during 1992-2010 period.

Table 2: Detail of landuse/landcover variations during 1992-2010 period

Landuse	1992		2000		2010		1992-2010	
	Area(ha)	%	Area(ha)	%	Area(ha)	%	Change (ha)	Change %
Conifer	1006	3.7	762	2.8	626	2.3	-381	-1.4
Scrub	15069	55.4	12485	45.9	10554	38.8	-4515	-16.6
Agriculture	1496	5.5	1958	7.2	2013	7.4	517	1.9
Rangeland	8024	29.5	9629	35.4	9982	36.7	1958	7.2
Soil/Rocks	326	1.2	870	3.2	1306	4.8	979	3.6
Settlement	762	2.8	1170	4.3	2421	8.9	1659	6.1
Water	517	1.9	326	1.2	299	1.1	-218	-0.8
Total	27200	100	27200	100	27200	100	-	-

Figure 5: Growth of urbanization is causing rapid landuse change in the Rawal watershed area.

The changes in landuse/landcover were variable on different elevation ranges during 1992-2010 period. The conifer forest has shown a decrease from 134 ha to 102 ha at greater than 1600m elevation range while this was from 343 ha to 238 ha within 1200-1600m elevation range during 1992-2010. The scrub class indicated a decrease of about 11 percent within 800-1200m range while 65% in less than 800m elevation range. In contrary to this, agriculture class had shown a increase of about 65% within 800-1200m range while 29%

increase in less than 800m elevation range. About 86% settlement class was found below 800 m elevation during year 2010 indicating most of the urban development in the low lying areas of the watershed.

Model Simulation

The model simulated an average water yield of about 378.6 mm/yr using base landuse of 2010 in the watershed area. About 49% of the yield was contributed by surface runoff and the rest by groundwater in the form of sub surface flows and springs etc. More than 70% of the annual yield was contributed during months of July, August and September. The surface runoff was found higher in the month of August i.e. over 83 mm while it was about 51 mm during July and 31 mm in September. The runoff was dominant over lower sub-basins likely due to higher impervious cover here than in the upper sub-basins of the watershed.

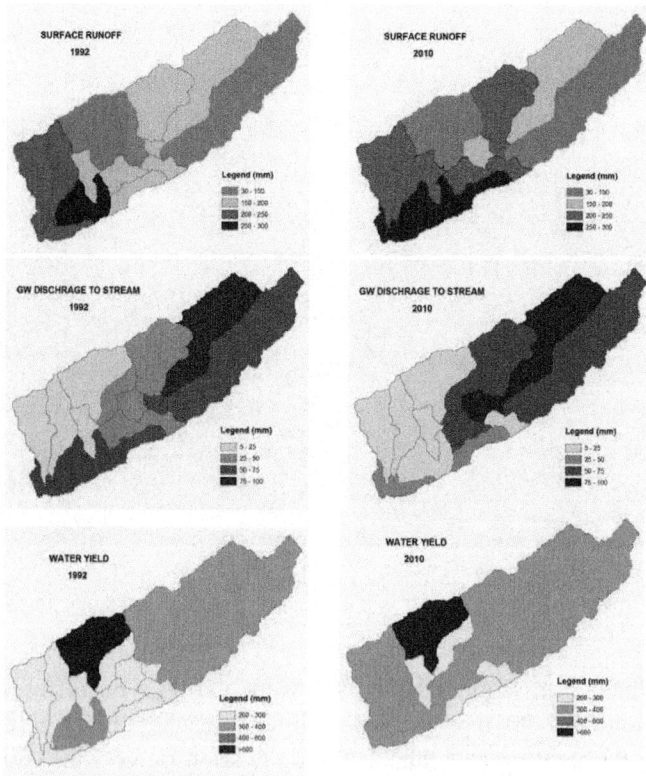

Figure 6: Comparison of the hydrological response of Rawal watershed to landuse conditions of 1992 and 2010 indicates dominant impact of landuse changes (i.e. urban development) in the southern low lying sub-basins on various hydrological parameters.

The groundwater discharge to stream flows was maximum in the month of September and more than 70% of the discharge occurred during period from August to December. The long-term average soil loss in the watershed was estimated over 17 tons $ha^{-1} y^{-1}$ i.e. ranging from 0.4 to 36 tons $ha^{-1} y^{-1}$ in different sub-basins. These estimates of soil loss matched closely with the results of [22] which indicated soil loss ranged from 0.1 to 28 tons $ha^{-1} y^{-1}$ averaging 19.1 tons $ha^{-1} y^{-1}$ at Satrameel study site in this watershed.

The model simulations showed a strong correlation between landuse evolution and the watershed runoff at the outlet. The change in landuse between years 1992 and 2010 indicated an increase of about 6% in the water yield and 14.3% in the surface runoff. The sub-basin wise hydrological response of the watershed during 1992-2010 period is shown in Figure 6. The sub-basins in the southern valley plains of the watershed indicate increase in surface runoff and water yield while decrease in groundwater contribution to the streams. The situation shows higher influence of urban landuse on hydrology of the low lying sub-basins as compared with sub-basins at higher elevations in the northeast of the watershed. Hydrologic changes due to increased impervious area and soil compaction generally lead to increased direct runoff, decreased groundwater recharge, and increased flooding, among other problems [27]. The combined effect of landuse and hydrological variations had exaggerated the problem of soil erosion resulting in an increase of about 17.4% in the sediment yield of watershed during 1992-2010 period.

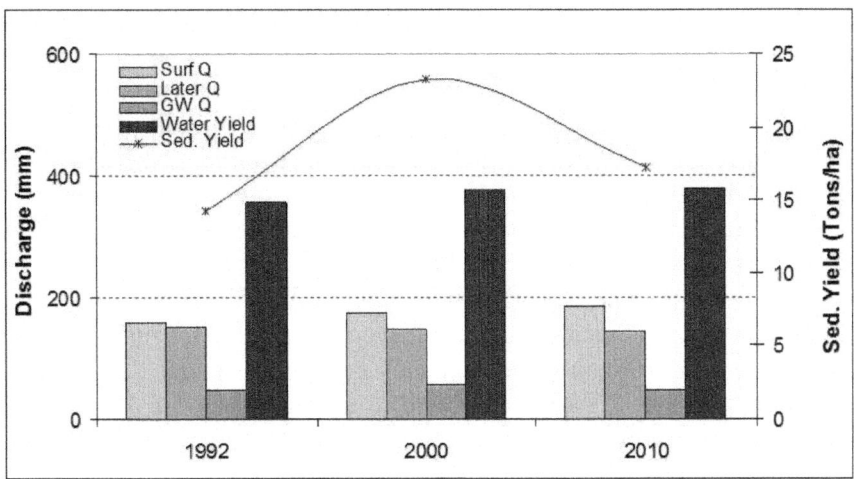

Figure 7: Hydrological parameters like surface runoff, water yield and sediment yield indicate an overall rise in values in reponse to landuse changes occurred within 1992-2010 period.

The increase in sediment yield can be attributed to the increase in surface runoff condition during this period (Figure 7). The zone of low sediment yield i.e. <5 tons ha^{-1}y^{-1} has shown a significant decrease while zones of medium sediment yield i.e. 5-10 tons ha^{-1}y^{-1} and high sediment yield i.e. >15 tons ha^{-1}y^{-1} a relative increase in the southeastern sub-basins of the Rawal watershed during 1992-2010 period (Figure 8).

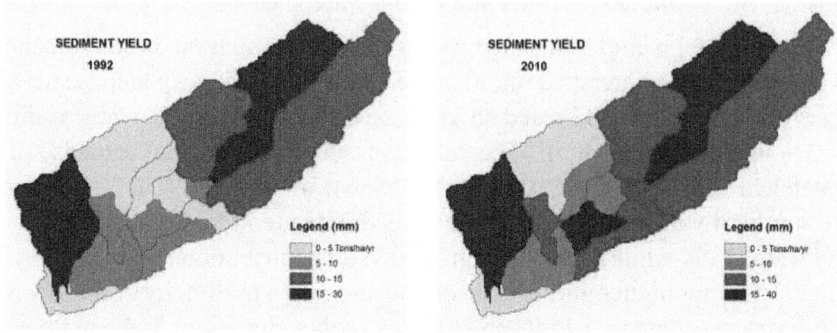

Figure 8: Temporal analysis of avarage annual sediment yield in the watershed during 1992-2010 peiod.

Scenarios of Extreme Conditions

Different scenarios of landuse and climate change were developed to observe the response of water and sediment yields to the expected extreme conditions in future. The first three scenarios are related to probable changes in landuse/landcover of the watershed in future. As most of the landuse changes are taking place due to growth in urbanization i.e. development of built-up, agriculture land, deforestation/ afforestation in the area, so it formed the basis of these scenarios. The percentage coverage of landuse in the watershed under base line and three scenarios is shown in Table 3 and in map form in Figure 9. The other scenarios are based on the prediction scenarios of climate change for this region i.e. +0.9 °C and +1.8 °C change in temperature during 2020 and 2050 [28] and changes in precipitation. These were formulated in consultation with experts from the Intergovernmental Panel on Climate Change (IPCC) and are consistent with the scenarios generated using the Model for Assessment of Greenhouse gas Induced Climate Change (MAGICC) software. The analysis of different scenarios is given below:

- In the first scenario, all the rangeland below 800m elevation is assumed to be converted into built-up land (About 20% increase in the settlement class). It is based on our study findings that most of the urban development has been occurred in the low valley areas below 800m elevation during the last two decades. The runoff estimates in urban areas are required for comprehensive management analysis. The scenario indicates a decrease of about 0.1% in the water yield while an increase of about 12.1% in the sediment yield from that of the base year 2010 (Table 4). The surface runoff has shown an increase of about 0.2% while lateral discharge a decrease of about 2% due to increase in the impervious area during 2010 in the watershed.

- In the second scenario, all the scrub forest below 1200m elevation is assumed to be converted into agriculture land (About 31% increase in agriculture land) keeping other landuse conditions same as of base year 2010. This scenario is also based on our study findings that major agriculture development has occurred below 1200 m elevation during the last two decades in the watershed area. The scenario indicates an increase of about 3.6% in the water yield and about 73.6% in the sediment yield from the base year 2010. The surface runoff increases by 4.4% while lateral discharge decreases by 5.6% due to decrease in the scrub forest coverage.

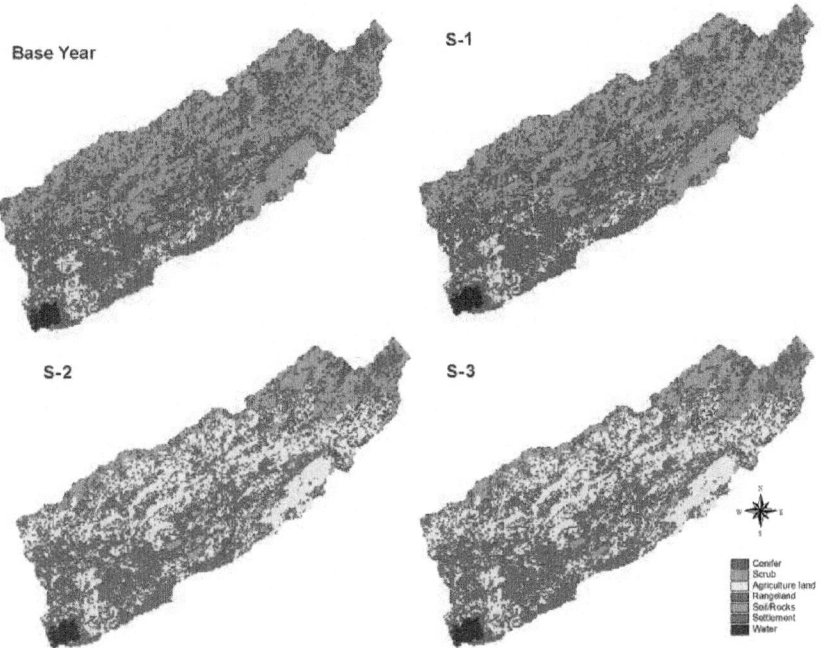

Figure 9: Landuse/landcover status during base year and three landuse change scenarios.

Table 3: Percentage coverage of landuse in base year 2010 and under three landuse change scenarios

Landuse	Base year 2010	Senario-1	Senario-2	Senario-3
Conifer	2.3	2.3	2.3	2.3
Scrub	38.8	38.8	7.7	7.7
Agriculture	7.4	7.4	38.5	38.5
Rangeland	36.7	16.3	36.7	16.3
Soil/Rocks	4.8	4.8	4.8	4.8
Settlement	8.9	29.3	8.9	29.3
Water	1.1	1.1	1.1	1.1
Total	100.0	100.0	100.0	100.0

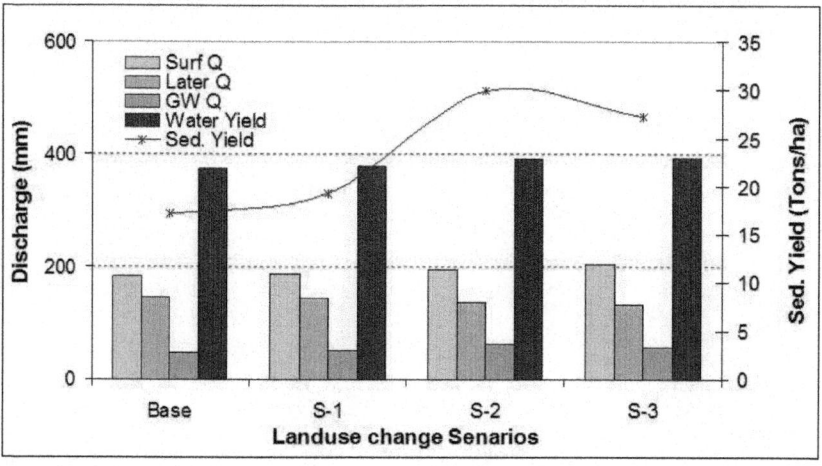

Figure 10: Hydrological response of the watershed under base year and three landuse change senarios.

Table 4: Percentage changes projected for surface runoff, water yield & sediment yield under different landuse and climate change scenarios using base conditions of 2010

No.	Scenarios	Surf. Runoff %	Water Yield %	Sediment Yield %
S-1	Urban develop. Below 800m elevation (converting rangeland into built-up land)	0.2	-0.1	12.1
S-2	Agriculture develop. below 1200m elevation (converting scrub forest into agriculture land)	4.4	3.6	73.6
S-3	Combining scenarios 1 & 2	10.5	4.1	58.1
S-4	+0.9°C temperature in 2020	-0.8	-1.3	13.1
S-5	+1.8°C temperature in 2050	-2.1	-3.0	28.3
S-6	Increase of 10% rainfall in 2020 with no change in temperature	24.5	19.1	26.1
S-7	Increase of 0.9°C temperature & 10% rainfall in 2020	23.6	17.8	41.8
S-8	Increase of 1.8°C temperature & 10% rainfall in 2050	22.1	15.8	58.1
S-9	Increase of 0.9°C temperature & decrease of 10% rainfall in 2020	-22.7	-19.0	-13.6
S-10	Increase of 1.8°C temperature & decrease of 10% rainfall in 2050	-23.8	-20.6	-1.5

- The third scenario is based on the combination of the 1st and 2nd scenarios i.e. increase in built-up and agriculture land below 800m and 1200m elevations, respectively. The scenario indicates an increase of about 4.1% in the water yield and 58.1% in the sediment yield of watershed. The surface runoff indicates an increase of about 10.5% while lateral discharge decrease of about 9% due to decline in the scrub forest cover and growth in the urban development. The scenarios 2&3 have also indicated an increase of 1.1% and 0.6% in actual evapotranspiration due to temperature variations. The hydrological response against different landuse change scenarios is shown graphically in Figure 10.

- The scenarios 4 to 10 are based on future climate changes in the watershed area and respective response of the water and sediment yields with the assumption that no change in landuse/landcover of base year 2010 will take place over the time. The rise of about 0.9°C temperature in year 2020 and 1.8°C in year 2050 indicates decrease of about 1.3% and 3.0% in the water yields of the watershed. Increase in temperature may result in higher evaporation rates that would affect the behavior of water yield.

- In scenario-6, increase of 10% in rainfall during 2020 keeping same temperature conditions as of base year 2010 has shown an increase of about 19% in the water yield and 26% in the sediment yield of the watershed. Similar increase in rainfall with same temperature conditions as of scenarios-4 and 5 i.e. +0.9°C in 2020 & +1.8°C in 2050, projects nearly 18% and 16% increase in the water yields and about 24% and 22% increase in the surface runoff as shown under scenarios-7 and 8 in Table 4. The sediment yield has also shown an increase ranging between 41% and 59% in these scenarios. The increase in rainfall usually causes increase in magnitude of floods which ultimately creates soil erosion and land degradation problems.

- On the other hand decrease of 10% in rainfall with same temperature conditions as of scenarios-4 and 5 projects decrease of about 19% and 21% in the water yields which ultimately lowered the sediment yield from that of base condition of 2010.

Risk Mitigation of Sediment Yield

Appropriate strategies have to be defined separately for different landuse conditions for minimizing the risk of soil loss and sediment yield involved in scenarios of extreme conditions. In order to reduce high sediment yield from sub-basins with rapid urban development, the unplanned urbanization needs to be controlled by appropriate laws and means. In the non-urban areas, proper soil and water conservation measures can be adopted to mitigate the risk of soil erosion.

In high risk zone of sediment yield, mainly dissected gullies are more susceptible to soil erosion. The problem of gully erosion can be solved to great extent through restoration of vegetative cover for which proper structures could be placed to provide protection long enough to give vegetation a start. The structures may be of temporary or permanent in nature keeping in view the nature of problem. Conservation structures to reduce velocity of the runoff can also be developed to control the extent of gully erosion. In medium risk zone of soil erosion, modifying the cross section and grade of channel to limit the flow velocities can be performed to stabilize the gullies. The conservation measures like terracing, contour bunding and diversion channels can be adopted in the contributory watershed to control excessive surface runoff causing gully erosion. These practices will also provide additional moisture for growing crops and vegetative cover thus help reducing gully erosion. In low risk zone of soil erosion, contour benches having small bunds crossway the slope of the land on a contour may be established to reduce the erosion risk. High intensity rainfall during monsoon season invariably cause over saturation harmful for plants. The situation can be avoided through provision of water ways and grassy outlets to dispose off the excessive runoff. The risk of erosion can also be minimized through adopting practices like strip farming, terracing, contour farming besides modifying bunds and minor land leveling in the cultivated area of watershed.

CONCLUSIONS

The recent changes in landuse/landcover conditions have brought significant impact on water flows, sediments and threat to eco-hydrology of the Himalayan area. The rapid growth in urbanization has increased the demand for land for development purposes consequently forest and water resources are coming under enormous pressure. The general trends of landuse change are gradual decline in coverage of scrub and coniferous forest, increase in urban development and somewhere in agriculture area. The increase in built-up land in the valleys has reduced the recharge source of groundwater which needs to be protected through controlling unplanned growth of urbanization. The rise in global warming accompanied with high variability in precipitation projects extreme changes in water balance and ultimately deterioration of the land quality. It is essential to regulate the urban development properly, affordable substitute-fuels for household use should be made available and an extensive community reforestation programme is undertaken to improve the fragile eco-system of the region. An integrated adaptation strategy needs to be developed at national and regional levels to cope with future implications of hydrological changes through focusing key policy areas and improving adaptive capacities of the communities at risk. Existing knowledge and data gaps need to be filled by systematic observations and enhanced capacities for research since these will be fundamental for developing climate change adaptation and mitigation programmes for the Himalayan region in future.

REFERENCES

1.   Behrman, N. The Waters of The Third Pole: Sources of Threat, Sources of Survival: 2010; 48. www.chinadialogue.net/UserFiles/File/third_pole_full_report.pdf (accessed 17 October, 2012).

2.   ICIMOD. Climate Change and the Himalayas: More Vulnerable Livelihoods, Erratic Climate Shifts for the Region and the World. Newsletter, Sustainable Mountain Development 2007; No.53: p55. ISSN 1013-7386 2007.

3.   Singh, SP; Bassignana-Khadka, I; Karky, BS and Sharma. Climate change in the Hindu Kush-Himalayas: The state of current knowledge, Kathmandu: ICIMOD; 2011.

4. Neitsch, S.L.; Arnold, J.G.; Kiniry, J.R.; and Williams, J.R. Soil and water assessment tool theoretical documentation. available at: http://swatmodel.tamu.edu/media/1292/ SWAT2005theory.pdf; 2005.

5. Neitsch S.L.; Arnold J.G. and Williams J.R. Soil and Water Assessment Tool, User's Manual; 1999.

6. Arnold, J.G.; J.R. Williams; R. Srinivasan; K.W. King and R.H. Griggs. SWAT - Soil and Water Assessment Tool, USDA, Agricultural Research Service, Grassland, Soil and Water Research Laboratory, 808 East Blackland Road, Temple, TX 76502, revised 10/25/94; 1994.

7. Williams, J.R. Chapter 25. The EPIC Model. p. 909-1000. In Computer Models of Watershed Hydrology. Water Resources Publications. Highlands Ranch, CO.; 1995.

8. Ndomba, P.M. Modelling of Sediment Upstream of Nyumba Ya Mungu Reservoir in Pangani River Basin, Nile Basin Water Science & Engineering Journal 2010; Vol. 3, No. 2: 25-38.

9. Setegen, S.G.; R. Srinivasan and B. Dargahi. Hydro-logical Modelling in Lake Tana Basin, Ethiopia Using SWAT Model, The Open Hydrology Journal 2008; Vol. 2: 49-62.

10. Odira, P.M.A.; M.O. Nyadawa; B. Okello; N.A. Juma and J.P.O. Obiero. Impact of land use/cover dynamics on stream flow: A case study of Nzoia River Catchment, Kenya, Nile Water Science and Engineering Journal 2010; Vol. 3, No. 2: 64-78.

11. Aftab, N. Haphazard colonies polluting Rawal Lake, Daily Times Monday, March 01. http://www.dailytimes.com.pk/default.asp...009_pg11_1. 2010.

12. Roohi, R.; Ashraf, A.; Naz, R.; Hussain, S.A. and Chaudhry, M.H. Inventory of glaciers and glacial lake outburst floods (GLOFs) affected by global warming in the mountains of Himalayan region, Indus Basin, Pakistan Himalaya. Report prepared for ICIMOD, Kathmandu, Nepal; 2005.

13. Ghumman, A.R. Assessment of water quality of Rawal Lake by long term monitoring. Environmental Monitoring Assessment Journal. DOI 10.1007/s10661-010-1776-x.; 2010.

14. Khan, M.I.R. Pakistan Journal of Forestry 1962; 12: 185.

15. Soil Survey Report. Reconnaissance soil survey of Haro Basin 1976, Preliminary Ed. Soil Survey of Pakistan, Lahore; 1978.

16. IUCN. Rapid environmental appraisal of developments in and around Murree Hills, IUCN Pakistan; 2005.

17. Tanvir, A.; B. Shahbaz and A. Suleri (2006), Analysis of myths and realities of deforestation in northwest Pakistan: implications for forestry extension. International Journal of Agriculture & Biology.1560–8530/2006/08–1–107–110. http://www.fspublishers.org.

18. Shafiq, M.; S. Ahmed; A. Nasir; M.Z. Ikram; M. Aslam and M. Khan. Surface runoff from degraded scrub forest watershed under high rainfall zone. Journal of Engineering and Applied Sciences 1997; 16(1): 7-12.

19. EPA (2004), Report on Rawal lake catchment area monitoring operation, Pakistan Environmental Protection Agency, Ministry of Environment, Islamabad: pp. 19.

20. Roohi, R.; A. Ashraf and S. Ahmed. Identification of landuse and vegetation types in Fatehjang area, using LANDSAT-TM data. Quarterly Science Vision 2004; 9(1): 81-88.

21. Oweis, T. and M. Ashraf (eds). Assessment and options for improved productivity and sustainability of natural resources in Dhrabi watershed Pakistan, ICARDA, Aleppo, Syria; 2012.

22. Nasir A.; K. Uchida and M. Ashraf. Estimation of soil erosion by using RUSLE and GIS for small mountainous watersheds. Pakistan Journal of Water Resources 2006; 10(1): 11-21.

23. Donigian Jr., A. S. Watershed model calibration and validation: The HSPF experience. National TMDL Science and Policy, Phoenix, AZ. November 13–16, 2002: 44-73.

24. Nash, J.E. and Sutcliffe V. River flow forecasting through conceptual models, I. A discussion of principles. Journal of Hydrology 1970; 10: 282-290.

25. Arfan, M. Spatio-temporal Modeling of Urbanization and its Effects on Periurban Land Use System. M.Sc Thesis. Department of Plant Sciences. Quaid-e-Azam University, Islamabad; 2008.

26. FAO. State of the world's forests – 2005. Food and Agricultural Organization (FAO), Rome, Italy; 2005.

27. Booth, D. Urbanization and the natural drainage system - impacts, solutions and prognoses. Northwest Environmental Journal 1991; 7: 93-118.

28. INC Report. Pakistan's Initial National Communication on Climate Change to UNFCCC, Ministry of Environment, Islamabad; 2003.

Chapter 10

IMPACT OF DROUGHT AND LAND – USE CHANGES ON SURFACE – WATER QUALITY AND QUANTITY: THE SAHELIAN PARADOX

Luc Descroix[1], Ibrahim Bouzou Moussa[2], Pierre Genthon[3], Daniel Sighomnou[4], Gil Mahé[3], Ibrahim Mamadou[5], Jean-Pierre Vandervaere[6], Emmanuèle Gautier[7], Oumarou Faran Maiga[2], Jean-Louis Rajot[8], Moussa Malam Abdou[1], Nadine Dessay[9], Aghali Ingatan[2], Ibrahim Noma[2], Kadidiatou Souley Yéro[1], Harouna Karambiri[10], Rasmus Fensholt[11], Jean Albergel[12] and Jean-Claude Olivry[13]

[1]IRD / UJF, Grenoble, France

[2]UAM University, Niamey, Niger

[3]IRD-HSM, Montpellier, France

[4]Niger Basin Authority, Niamey, Niger

[5]University of Zinder, Niger

[6]UJF-LTHE, Grenoble, France

[7]Université Paris 8, France

[8] IRD-BIOEMCO, Créteil, France

[9]IRD-ESPACE-DEV, Montpellier, France

[10]2iE International high School, Ouagadougou, Burkina Faso

[11]University of Copenhague, Denmark

[12]IRD-LISAH, Montpellier, France

[13]IRD, France

INTRODUCTION

West Africa has been experiencing drought conditions since the end of the 1960s. This pattern has been particularly evident in the Sahel, but appears to have attenuated in the last decade in the eastern and central parts of this region.

On the other hand, annual rainfall remains very low in the western part of the Sahel [1].

A corresponding decrease has also been observed in the mean annual discharge of the Senegal and Niger rivers, which are the largest in the region and primarily fed by water originating from tropical humid areas. However the percentage decrease in mean annual discharge was almost twice as large as the decrease in rainfall [2] for the period 1970-2010. Similar trends have been observed on smaller river systems.

In contrast, even though the Sahel and most of West Africa also have experienced substantial drought over the past 40 years, runoff coefficients and stream flows have increased in most Sahelian areas. This phenomenon has been named "The Sahelian Paradox" after the increase of the groundwater table in Niger since the 1960s was named the Niamey paradox and attributed to substantial changes in land-use. The HAPEX-Sahel (Hydrological and Atmospheric Pilot Experiment) and the AMMA (African Monsoon Multidisciplinary Analysis) programs have provided, among many comprehensive results, valuable measurements dealing with the spatial and temporal variations in Sahelian soil water content as well as with the infiltration of water through deep soil layers of the vadose zone.

The purpose of this chapter is to provide an overview of hydrological behaviour throughout West Africa based on point, local, meso and regional scales observations.

BACKGROUND

The paradoxical increase in runoff despite drought conditions in sub-Saharan Africa was first noted in a paper by Albergel [3], analysing decadal series of runoff measurements in experimental sites of Burkina Faso. He noticed that this increase was observed in Sahelian areas, but not in the more humid Sudanian regions.

"The decrease in rainfall during the 1969-1983 period seems to be largely offset by the evolution of surface features in the functioning of small catchments. These changes favoured the conditions of runoff in the Sahelian basins; there are due to both the human actions and the climatic conditions. The reduction of vegetation cover and the widespread crops areas cause soil surface settling and the appearance of impervious superficial layers, as well as the extension of eroded areas. Some sahelian basins have nowadays [in 1987] the common characteristics of basins located northward, with great areas of bare soils; perennial graminaceae are replaced with annual ones, and combretaceae with prickly bush species" [3].

Albergel [3] attributed the contrasting behaviour of Sudanian (mean annual rainfall > 750 mm) and Sahelian (mean annual rainfall < 750 mm) areas to increasing bare soils and decreasing vegetation cover in Sahelian basins.

This hypothesis was confirmed in 1999 by Mahé and Olivry [4] and then in 2002 by Olivry [2], who remarked that the discharge of right bank tributaries of Middle Niger River had been increasing since the beginning of the Drought (1968). Similarly, Amani and Nguetora [5] noted that runoff coefficients were increasing significantly in right bank tributaries and showed that the onset of the annual flood was occurring earlier than in previous decades.

Mahé *et al.* [6] analysed the runoff evolution of eight right bank tributaries of the Middle Niger River and noted that the decrease in rainfall did not lead to a decrease in runoff under the Sahelian climate as commonly observed in other basins in the world. Rather, these tributaries exhibited increasing runoff coefficients and in discharges, while "Sudanian" climate tributaries suffered a decrease in discharge and in runoff coefficient [6].

MATERIAL AND METHODS

This study is mainly based on two sources of data:

- field measurements and observations made during the AMMA (African Monsoon Multi-disciplinary Analysis) experiment at the Niger experimental site (Niger River middle stretch and Niamey square degree), and:

- rainfall and discharge data collected on the operational network of Niger basin, provided by the Met Offices of Niger, Mali and Burkina Faso and by NBA (Niger Basin Authority).

The methods included the following:

- Analyse of runoff and river discharge data (in order to characterize the trends in the river discharge records) at several scales:

  - At the point scale: infiltration tests (using disk infiltrometers at multiple suctions) and soil water content monitoring provides data on soil hydraulic conductivity and other physical properties [7] [8];

  - At the local scale: Tondi Kiboro and Wankama catchments, as well as 20 experimental plots of 10 and 100 m², located in the same catchments. These data were collected during the AMMA experiment (2004-2010). On the plots, the measurements were made after each event; on the catchment, stream gauges allowed the monitoring of the discharges[9];

- At the meso scale:

  - Some small direct tributaries of the Niger River; we overall use here the regional balance allowed by the stations located in the Niger River upstream (Kandadji) and downstream (Niamey) the studied stretch; however, some discharge data of small direct tributaries were collected for this study [10];

  - The main tributaries of Niger River's middle stretch.

  - At the regional scale: the Niger River, the Senegal River, etc, existing data, allowed these analyses [10] [11].

- Analyses of land cover data (including agricultural data, NDVI etc.) and a map of land cover in the square degree of Niamey were realised during AMMA experiment;
- Analyses of precipitation trends across the Sahel ;
- Analysis of endorheism breaks was carried out in the region of Niamey [11].

DECREASE IN RAINFALL, INCREASE IN RUNOFF

Increasing Streamflows

The Great Drought of the Sahel is considered one of the most significant climatic events worldwide [12]. For at least 25 years, more than 3 millions km² of semi-arid Sub-Saharan area has suffered a rainfall deficit ranging from 10 to 30%, depending on the location. Figure 1 shows a partial offset of the deficit since the mid-1990s. However, the overall deficit remains and the interannual variability has increased during this period. Not shown in this figure, the intensity of the drought has been largely attenuated in the eastern part of West Africa since the 2000s, but it persists in the western part of the region. In spite of the severe drought, a significant increase in the runoff coefficient and a general increase in stream discharge, have been observed (see background) since the 1960s, in the Sahelian basins (Figure 2).

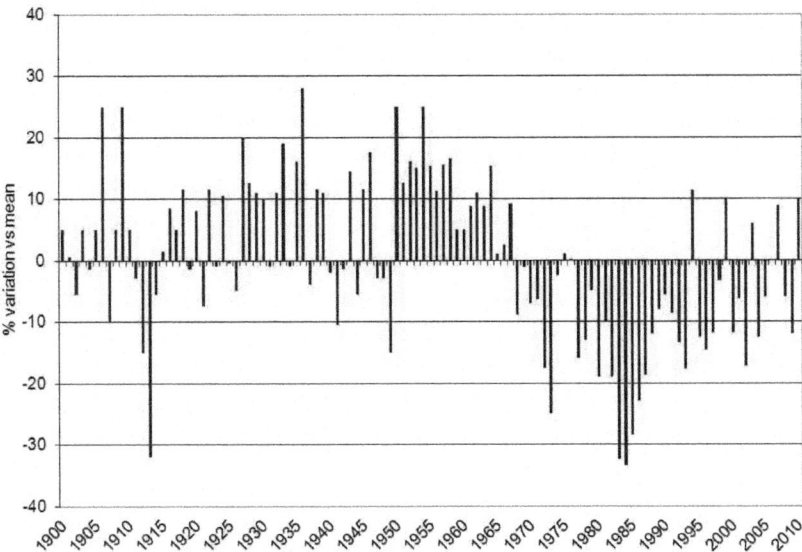

Figure 1: Evolution (1900-2010) of the Standard Precipitation Index for the whole Niger River basin.

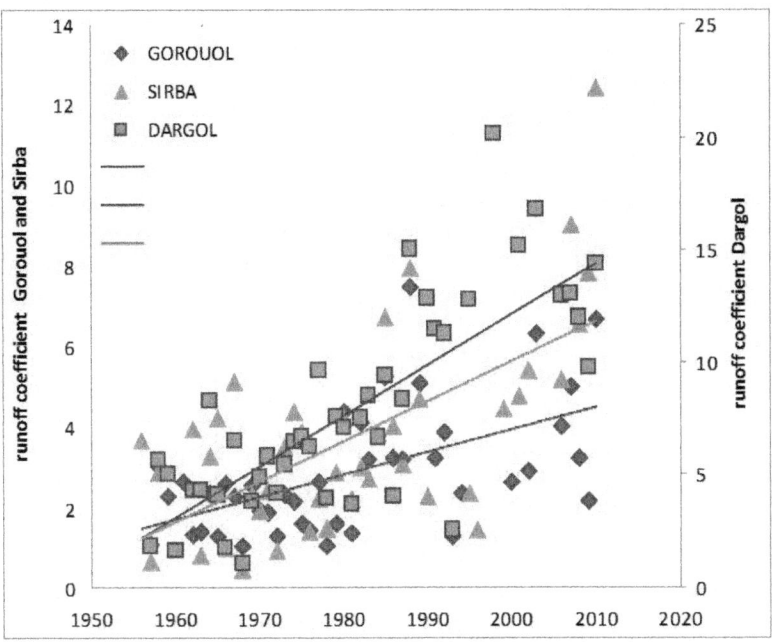

Figure 2: Evolution (1955-2010) of the runoff coefficient of three main right bank tributaries of Niger River.

Figure 3 shows the location of the middle Niger River basin in West Africa. Tributary rivers accounted here are located in the eastern blue circle representing the middle basin. Due to the great loop created by the Niger river northward to the margin of the Saharan desert, the annual flood downstream from the Niger Inner Delta -a large humid area located in the northern reach of the river-, in the Middle Niger River, has two flood peaks. The first one, termed the red or local flood, arises from local rainfall draining through a series of tributaries (Fig. 3, enlargement) and occurs between August and September. The second, termed the Guinean flood, originates from precipitation in the Fouta Djallon (Guinea) area during the rainy season (June–September). Delayed by the crossing of the Niger Inner Delta (see Fig. 3) in Malian territory, the Guinean flood takes place around January. The clear separation between these two flood events makes it possible to distinguish the local Sahelian effect from the more remote trend.

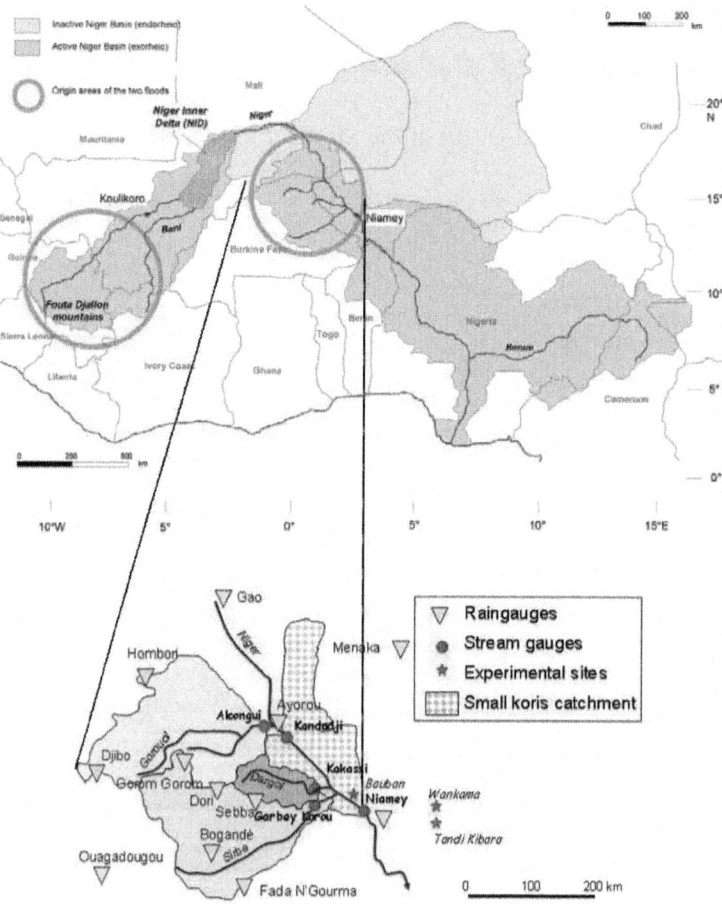

Figure 3: The Niger River basin and (enlargement) its middle basin.

Earlier Flooding

Another change observed during the drought is the earlier onset of yearly flooding, compared with previous periods. Figure 4 shows that the first flood is now arriving approximately forty days earlier than it did forty years ago. This observation is consistent with a decreased soil water holding capacity in the river basin.

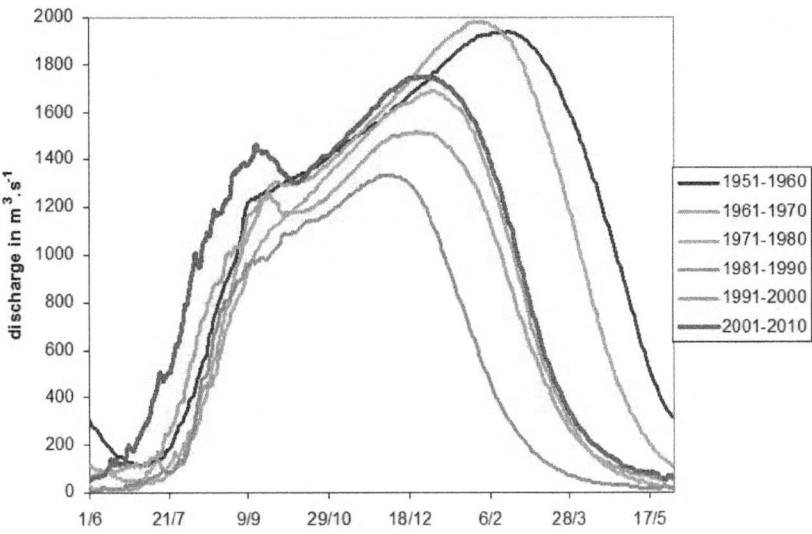

Figure 4: The Niger River decadal hydrographs at Niamey station (Republic of Niger).

Sahelian Paradox is Not Due To Rainfall

Because rainfall has decreased significantly since 1968, rainfall amount does not explain the increase in runoff and stream flow or the earlier flood occurrence. For example, the three main tributaries affected by the semi-arid Sahelian climate (The Gorouol, the Dargol and the Sirba rivers: see enlargement of figure 3) experienced a significant increase in runoff coefficient during the drought, despite a 20% reduction in rainfall (Table 1).

To identify the drivers of increased stream discharge and early flood onset, we analysed the trends and evolution of rainfall for twelve stations with daily rainfall data from 1950 onward. These stations are indicated in figure 3 [11].

A forward shift in the timing of the monsoon rains does also not appear to explain the Sahelian paradox. Nicholson [13] as well as Ali and Lebel [1] observed in recent decades a reduction of rainfall in August and a relative increase in rainfall amount in June and July. A similar forward shift in monsoon timing was observed in the Middle Niger River basin (figure 5). However the total amount of rainfall in June and July remains lower during the last two decades than during the 1950s and 1960s (figure 6).

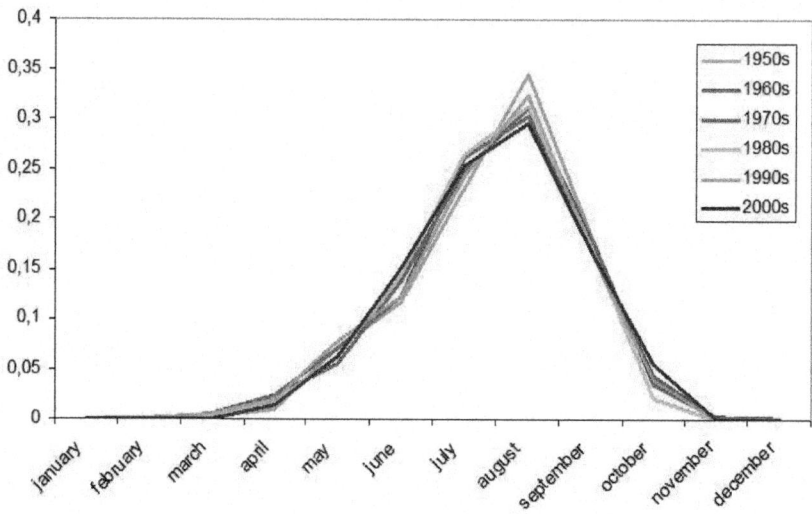

Figure 5: Evolution per decades of the rainfall monthly index (middle Niger River basin).

- Because the runoff coefficient increases with increasing rainfall intensity and amount, a rise in the number of extreme events might explain the higher volume of the floods. However a general decrease in the number of extreme rainfall events has been observed since the 1950s, for each class of events (figure 7), ranged by total amount of the event (classes 20-30 mm; 30-40 mm; 40-60 mm and more than 60 mm). However at the whole Sahel scale, a current study shows an increase in the rainfall amount for events upper than 30 mm during the 2000s.

- An increase in extreme events at the beginning of the rainy season also could explain the early flood occurrence. An increase in the total amount of rainfall fallen in events > 40 mm has been observed in the last decade, but only in June. However, the runoff coefficient is nowadays two or three times higher in the Middle Niger River basin than during the 1950s. Thus, the modest increase in rainfall amount (event > 40 mm) observed in June during the 2000s cannot alone explain the timing and magnitude of the recent floods.

Table 1: Evolution of runoff coefficients of the three main Niger River right bank tributaries from 1957

Basin	Gorouol	Sirba	Dargol
Period 1957-1979	1.9	2.6	5.0
Periods 1980-1994	3.6	4.2	8.8
Periods 1995-2010	4.3	6.0	14.2

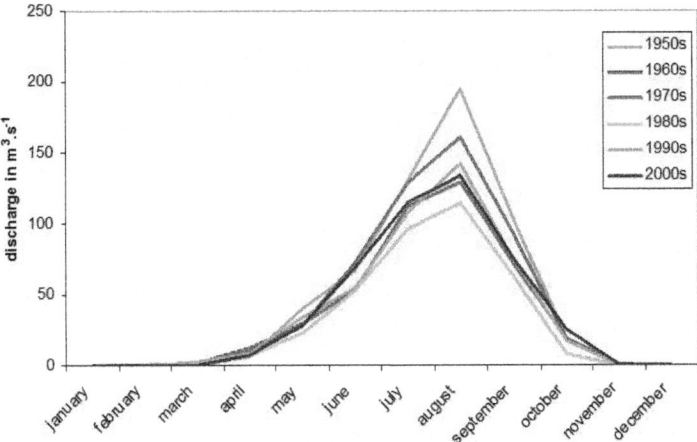

Figure 6: Evolution per decades of the monthly mean rainfall amount (middle Niger River basin).

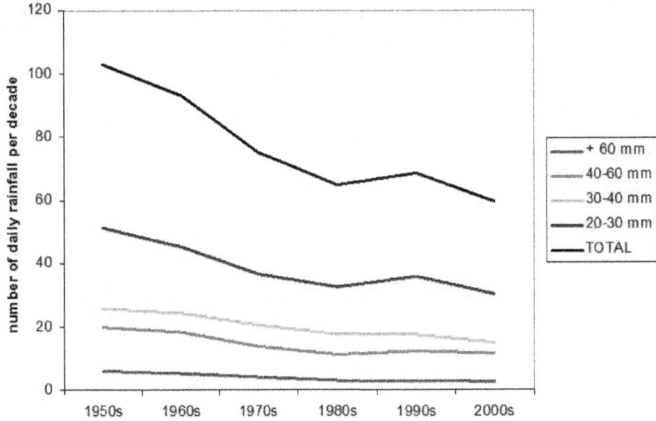

Figure 7: Evolution per decades of the number of extreme events (ranged by total rainfall amount of the rainy event), middle Niger River basin.

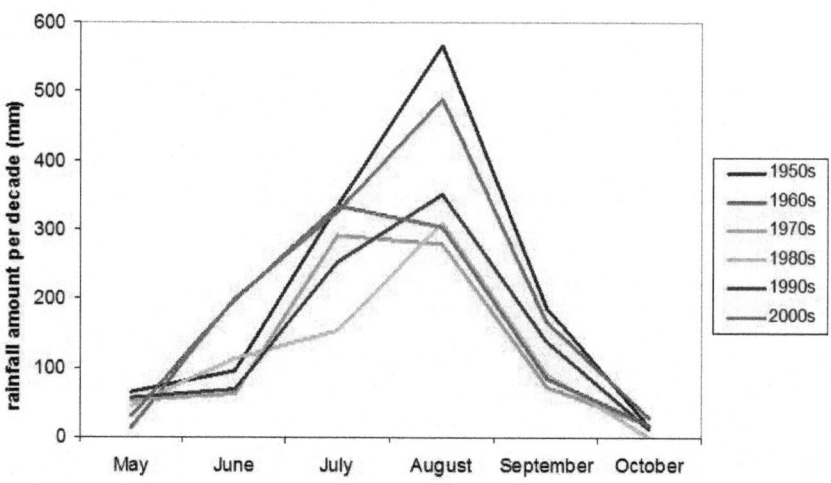

Figure 8: Evolution per decades of monthly rainfall amount of events > 40 mm Middle Niger River basin.

CONTRIBUTION OF CHANGING SOIL CHARACTERISTICS TO SAHELIAN PARADOX?

In Sahelian areas at all scales, runoff coefficient generally increased along with river discharges (Table 2). In most cases, these changes correlate with a decrease in vegetation cover, due to land use changes including increasing crop area, overgrazing and wood harvesting. At the regional scale, the change in vegetation cover is not obvious. The runoff coefficient of degraded soils in the Sahel (close to 60% for the ERO type crusted soils –as defined by Casenave and Valentin [14])- is much higher than that observed for millet crops (4%) and for bush and fallows (10%; see Table 1 and figure 9). Small crusted soil areas can alone explain increased stream flows. Therefore increasing runoff coefficients may be consistent with the re-greening documented in remote sensing studies (higher values of NDVI), the remaining soils being covered by higher millet crops density, graminaceae, herbaceae and annual plants than during previous periods.

Runoff coefficients are measured in plots of 10 and 100 m²; saturated hydraulic conductivity is measured in at least 20 points for each surfaces class [11].

Figure 9: Opposition between a "fallow on common sandy soil" (left) plot and a "fallow erosion (ERO) crusted soil" plot (right).

Table 2: Comparison of the hydrodynamic properties of non-crusted and crusted soils

Soil surface feature	Runoff coefficient %	saturated hydraulic conductivity mm.h-1
Millet on common sandy soil	4.0 +/- 1.4	172 +/- 79 (20)*
Fallow on common sandy soil	10 +/- 4	79 +/- 41 (20)
Old fallow with bioderm	25 +/- 7	18 +/- 12 (30)
Millet and fallow erosion (ERO) crusted soils	60 +/- 8	10 +/- 5 (30)

[i] - number of repetitions

Table 3: Runoff characteristics for the periods 1 (1991-1994) and 2 (2004-2009) on the three small Tondi Kiboro basins

1991-1994	Eainfall	Runoff depth	Kr*	Rainfall/ runoff	r^2 R = a P + b	Yearly runoff total duration in hours
TK amont	513	180	0,36	R = 0,56 P - 2,61	0,82	34,9
TK aval	513	133	0,26	R = 0,43 P - 2,3	0,79	28,1
TK bodo	485	185	0,38	R = 0,53 P - 2,14	0,68	62,7

2004-2009	rainfall	Runoff depth	Kr*	Rain-fall/runoff	r² R = a P + b	Yearly runoff total duration in hours
TK amont	495	231	0,47	R = 0,77 P – 4,9	0,85	34,2
TK aval	491	132	0,27	R = 0,49 P – 3,5	0,74	18,2
TK bodo (2007-2009)	520	242	0,47	R = 0,87 P – 7	0,81	25,9

[i] - Kr = runoff coefficient

Hydrodynamic characteristics differ substantially between different types of soil surface features. Given the strong difference in runoff coefficients, land use /land cover evolution could probably explain the runoff increase and "re-greening" is evident in the Sahelian area (see section 9 below). However, what was rather observed within the AMMA experimental sites in Niger is a degradation of soils and vegetation during the last decades, without a noticeable recovery since the mid 1990s.

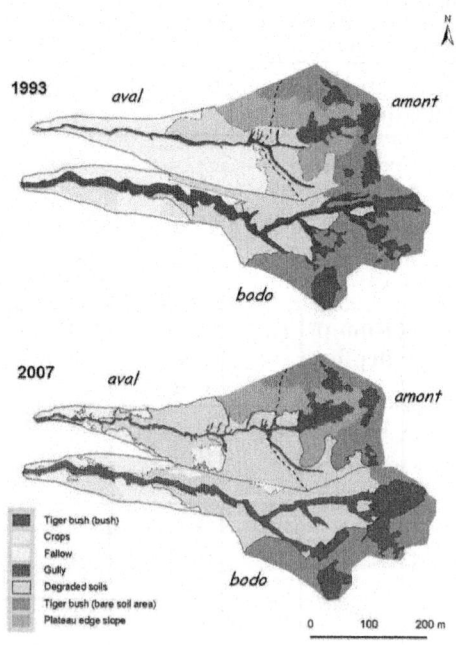

Figure 10: Land use over the Tondi Kiboro experimental catchments (Western Niger) in 1993 (up) and 2007 (down).

In the Tondi Kiboro catchments, the area of degraded soils (mostly ERO crusted soils) in 2007 was twice that observed in 1993 (figure 10). As a matter of fact, the runoff coefficients were significantly increasing from the 1991-1994 to the 2004-2009 period (Table 3), as a consequence of the reduction of soil water holding capacity and the rise in crusted surfaces.

In the Wankama experimental catchment, the same evolution of land cover is observed; but there is no historical hydrological data available for comparison. However, figure 11 allows comparing the vegetation cover in 1950 (aerial photos) and in 2007 (pictures taken from a PIXY® drone; [15]).

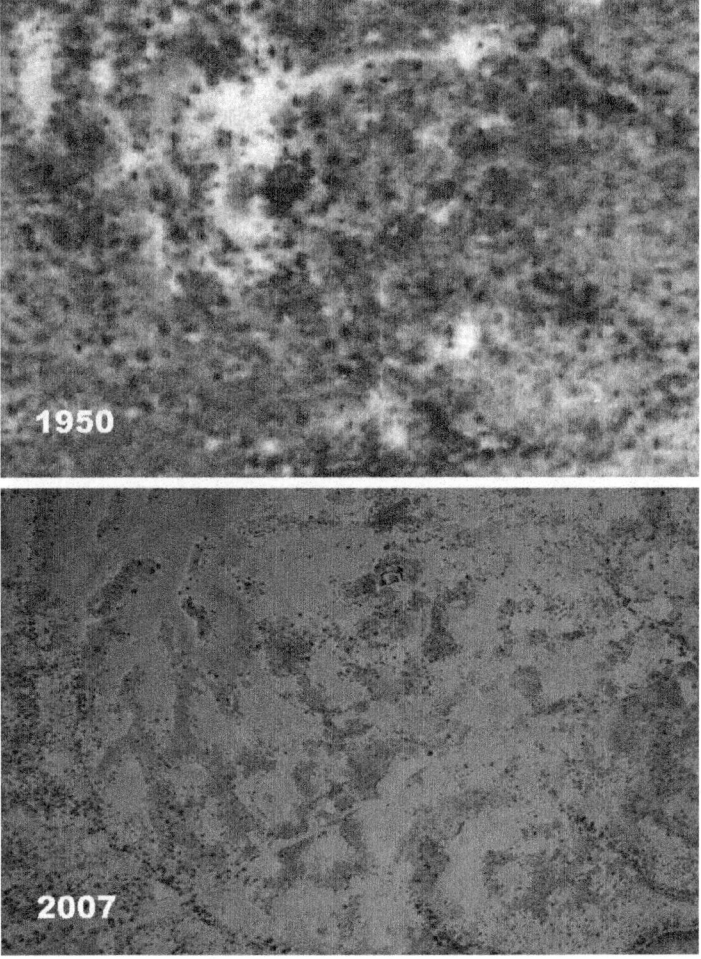

Figure 11: The lower part of Wankama experimental catchment in 1950 (up, aerial photo by IGN France) and in 2007 (photo taken from a PIXY drone); the generalisation of ERO crusted soils in a way of "hydro-aeolian depressions" is noticeable.

In the lower part of this basin, the "ERO" crust areas are widespread and constitute a very active contributing area; 70% of the surface is covered by "ERO" crust (see Fig. 9 at right hand side). The ERO crust runoff coefficient is approximately 60% (Table 2) while it is only 4% in pearl millet crops and 10% in the fallow. An increase in "ERO" crusted area must then have strong hydrological consequences.

The "small koris" catchments (see Fig.3) are not gauged. However, taking the difference between the discharge at Niamey station, on the other hand, the sum of discharges of Niger at Kandadji and those of the Sirba and the Dargol (the two tributaries feeding Niger river between Kandadji and Niamey, Fig.3)[11] showed that a clear change in the behaviour of these "small koris" behaviour occurred after 1997. In general, between 1975 and 1996 the input volume of the Niger River at the Kandadji station was greater than the output volume at Niamey, presumably due to infiltration and evaporation losses. Since 1998, the opposite is observed, presumably due to new input from small tributaries, where crusted soils areas have increased in recent years. Some of these basins have shifted from endorheic prior to and during the 1990s to exorheic in recent years (see 6. section below), increasing the contribution to the Niger River from degraded areas with high runoff. These small tributaries have recently provided several billions cubic meters per year to the Niger River discharge [11] (Fig.12).

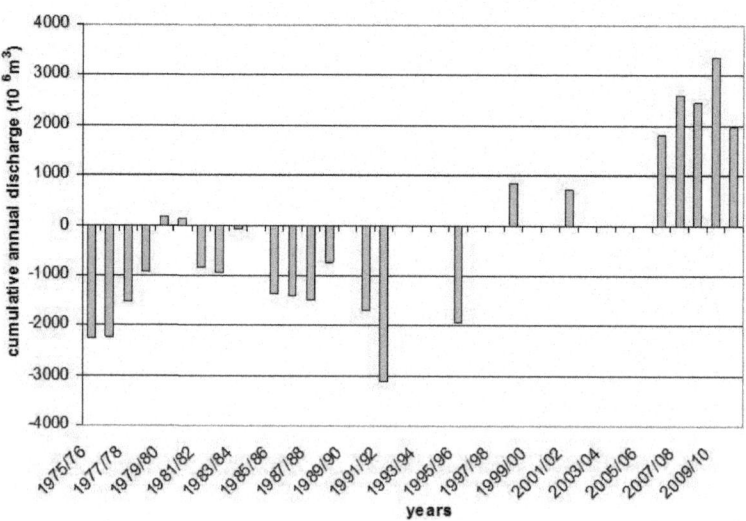

Figure 12: Evolution of the remaining water balance between Kandadji and Niamey from 1975 to 2010, after subtracting the discharge of the two main right bank tributar-

ies (Dargol and Sirba); discharge of the Dargol and Sirba rivers were measured at the Kakassi and Garbey Korou stations (see Fig. 3).

This cited study [11] was dedicated to the severe flooding of August and September 2010 in Niamey, were 300 ha in the river right bank were inundated. Twice, the level of Niger River reached a maximum during the rainy season (2030 $m^3.s^{-1}$ in early August, 2130 $m^3.s^{-1}$ in early September, the maximum previously registered value being 2000 $m^3.s^{-1}$). However, during August 2012, this maximum was widely exceeded. The discharge value reached 2473 $m^3.s^{-1}$ on 18th August, causing severe damage in the city of Niamey and extensive flooding downstream in Niger, Benin and Nigeria (Figure 13) [16].

Thus, previous studies show that from the measurement point scale to the meso-scale basin, the runoff increase is observable in the Sahelian area of the Middle Niger basin:

- at the point scale, it is shown that the new surface features created by land use change (crusted soils areas particularly) have a very low hydraulic conductivity and consequently high runoff coefficients (Table 2), the latter being measured in plots of 10 and 100 m^2;

- at the small basin scale, the experimental basins of Tondi Kiboro (12 ha) exhibited an increase in runoff and discharge probably due to the extension of crusted soil areas (Figure 10 and Table 3);

- the direct middle Niger river tributaries experienced a strong runoff increase after 1997 (Figure 12); the corresponding scale ranges from 10 to 2000 km^2;

- the meso-scale Niger right bank tributaries basins range from 7000 km^2 (Dargol river basin) to 38,500 km^2 (Sirba) and 45,000 km^2 (Gorouol); they have shown a strong discharge increase since the beginning of the drought (figure 2); Amani and Nguetora [5] demonstrated that the flood was occurring during the 1980s almost one month earlier than during the 1960s, as a consequence of both decreasing vegetation cover and reduction in soil water holding capacity;

- the previous statements explain the significant rise in the first, red flood of the Niger river in its Middle basin (figure 13) and the earlier occurrence of this first flood (40 days earlier than during the 1970s as seen in figure 4).

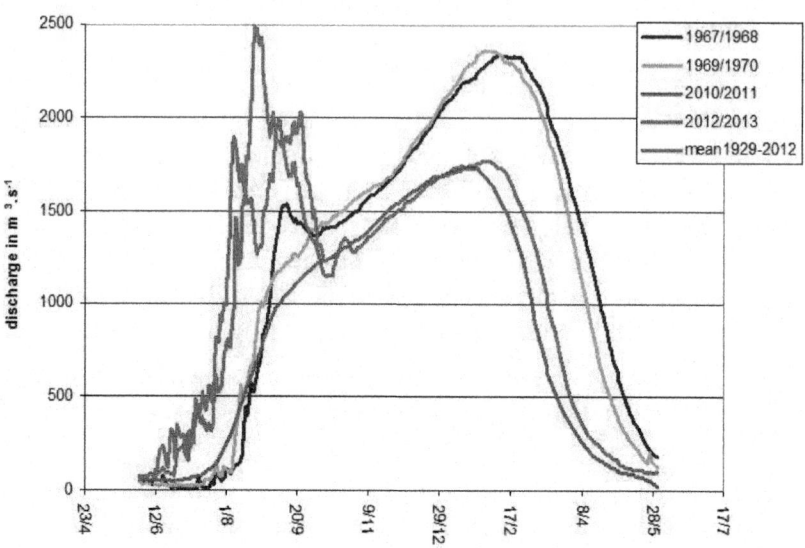

Figure 13: The 2012 red flood compared with previous remarkable years, included the two severe floods during the monsoon of 2010, and the pattern of the Guinean or black flood [16]

AN INCREASE IN FLOOD HAZARD

"As it has been supposed from the end of the 1980's, the change in the hydrological functioning and the newly twin peak hydrograph" (of the Middle Niger river) "are linked to human factors, mostly to land use changes, particularly the land clearing and the extension of crops due to demographic pressure; this led to a soil baring and a fallow shortening which caused a soil crusting resulting in a severe decrease of soil infiltrability. This results in an increase in flood hazard" [11].

The record high 2012 flood must remind policy makers that this hazard is becoming a big social and environmental concern to be accounted in land planning policy. Overall, environmental engineering must be performed and improved in order to offset and mitigate the effects of this trend: urban areas are firstly affected by flooding, but rural areas are those which have to be land managed in order to increase the soil water holding capacity. This should allow increasing land productivity and reducing the flood hazard downstream.

The immediate causes of the 2012 actual flood are not yet determined; however, rainfall amount was high, probably the highest measured since 1968 in the Middle Niger Basin. We focus here on the 2010 monsoon flood.

The first, small, mid July 2010 peak was due to all the sub-basins. It is worth noticing that the first of the two higher rainy season peaks (early August) was firstly due to discharges coming from Malian territory, upstream from Kandadji station and, then, to stream flows coming from the Gorouol basin. The second peak (early September) was mostly produced by flows coming from upstream from the Gorouol confluence, thus from the arid area of eastern Mali, and secondarily by the Sirba and Dargol rivers discharges. The small koris contribution is only estimated, as shown, by the difference between the discharge at Niamey and the sum of Niger at Kandadji, the Dargol and the Sirba. However it is obvious that these small koris had a large contribution to the first peak (figure 14).

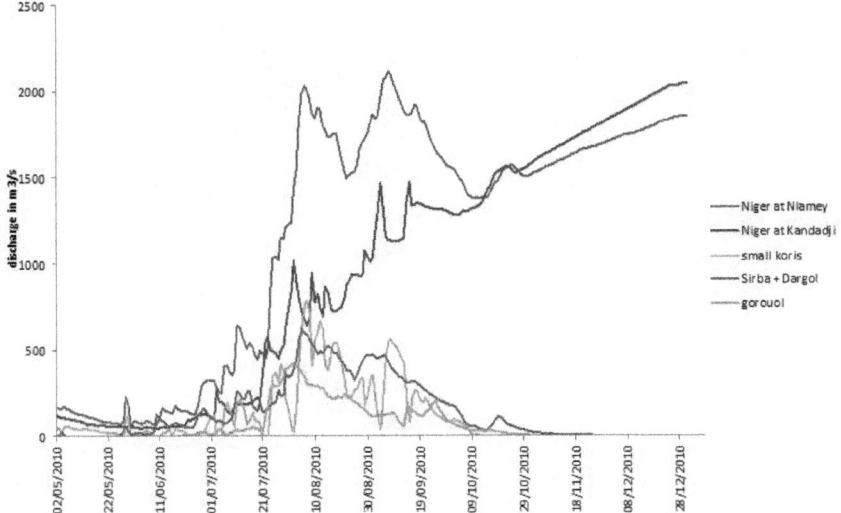

Figure 14: flood decomposition between different contributing areas.

The long term causes of these floods are the land use change and the increase in crusted soil areas. But recently, flood hazard was accentuated by three factors:

- the urbanisation
- the silting up
- the endorheic bursting

Urban population in Sub-Saharan Africa remains the lowest in the world. It is expected to increase strongly in the next years and decades. The percentage of urban population in the Sahel is only 30% in average (but 17% in Niger and 20% in Burkina Faso), compared with 50% in the southern West African countries on the shore of the Gulf of Guinea. The urban population in the Sahel is expected to reach 40% in 2030 and 50% in 2050. As the total population is increasing by 2.5-3% per year, it duplicates every 20-25 years. The urban population doubles every 14-15 years. Most of new resident come from rural areas and settle familial or informal housing. The latter is commonly settled on non-drained and non-buildable areas, which constitute the first areas being inundated in case of flooding. This was the case in 2003 at Saint Louis (Senegal), the 1st September 2009 at Ouagadougou (Burkina Faso) [17] or Agadès (Niger), in August 2010 and August 2012 at Niamey (Niger) [16], and at Dakar (Sénégal). Most of informal housing is built in adobe, straw or iron sheet, which is destroyed or severely damaged by flooding (Fig.15).

Figure 15: Flooded areas in Niamey (shore of Niger River) the 18th August 2012 (left) and in Agadès (right) the September 1st 2009 (*photo Ibrahim Noma and Baptiste Nay*).

In southern Sudanian areas [18] [19], the ongoing increase in the occurrence of inundations was highlighted during the last years. Flood related fatalities in Africa, as well as associated economic losses, have increased dramatically over the past half century [20]. This trend associated with urbanisation is expected to have dramatic consequences.

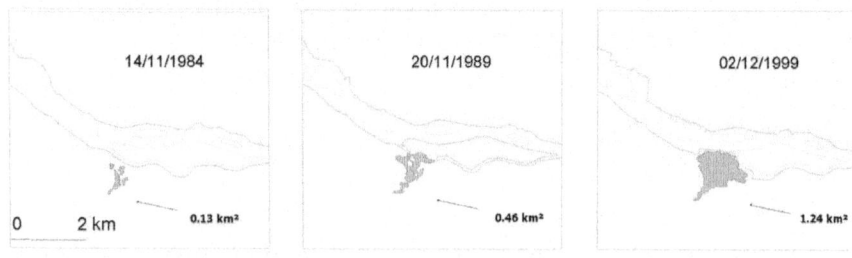

Figure 16: Evolution of the alluvial fan of the Kourtéré kori into the Niger river bed, just upstream from Niamey (Niger) [10].

The second increasing flood hazard is the current silting up of river beds in the Sahelian region, linked to the observed erosion stage [21]. As runoff is rising, erosion is exacerbated, increasing sediment transport, leading to sandy deposits on the river beds, and contributing to flooding. This was the case for the last Niger River floods in Niamey, due to the invasion of its bed by alluvial fans coming from tributaries upstream from Niamey (Figure 16).

A third increasing flood hazard in the Sahel is the increased exorheism. Some endorheic valleys became exorheic in recent years and decades, increasing the Niger river catchment area and the number of tributaries. Due to soil crusting, these new contributing areas have high runoff coefficients causing a rise in the available discharge for the Niger River (figure 17).

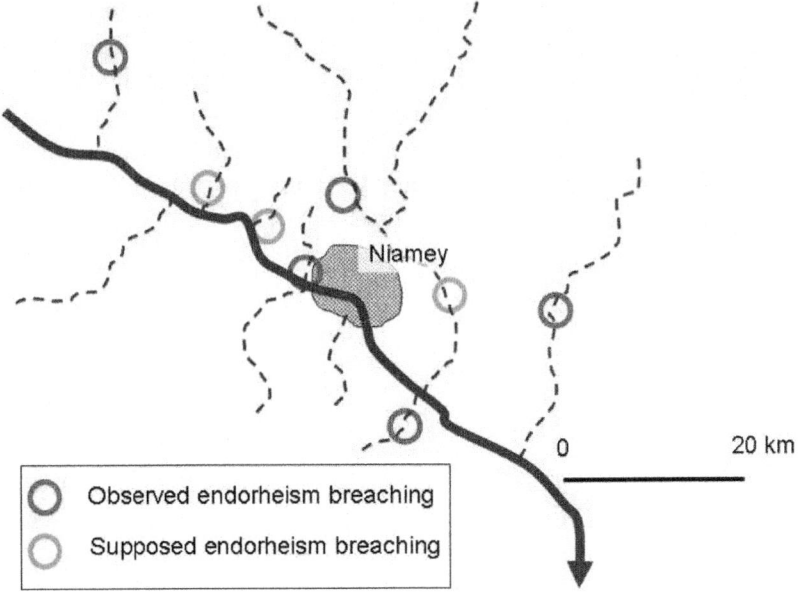

Figure 17: Observed and supposed endorheism bursting in the Niamey area of Niger River valley.

REGIONALISATION

The runoff increase is not observed only in the Middle Niger river basin. The important contributions of Gil Mahé [4] [6] [22] [23] [24], before a first attempt of synthesis of Amogu *et al.,* [10], show the large extension of the increase in rivers discharge, justifying the termed "Sahelian Paradox".

We propose here a regionalisation of such mechanisms and an analysis of the respective role of natural (climatic) and Human (land use changes) factors in the appearance, evolution and geographical distribution of these hydrological behaviours.

Mahé and his colleagues highlighted the rise in the Nakambé River (Burkina Faso) discharge ([22];figure 18), after documenting the same process on the Sahelian Niger river right bank tributaries [6]. More recently, they observed a similar evolution in the western part of the Sahel, in southern Mauritanian rivers ([24]; figure 19). In a newly published paper [23], the discharges in the Sokoto river in Nigeria showed an increasing trend, similar to those of the right bank tributaries.

The stream discharges of Sudanian rivers (Sudanian climate is characterized by annual rainfall amount exceeding 750 mm) were found to be decreasing [10], which was proposed by Mahé [25] to result from a drop in the water level in the aquifers sustaining the rivers during the low flow period. These areas have a "Hewlettian" hydrological behaviour instead of the "Hortonian" functioning typical for the Sahelian semiarid regions. Runoff only onset when the soil is saturated; in these areas, the reduction in rainfall affected firstly the part of rainy water previously dedicated to runoff, explaining the significant runoff and discharges decrease in these areas.

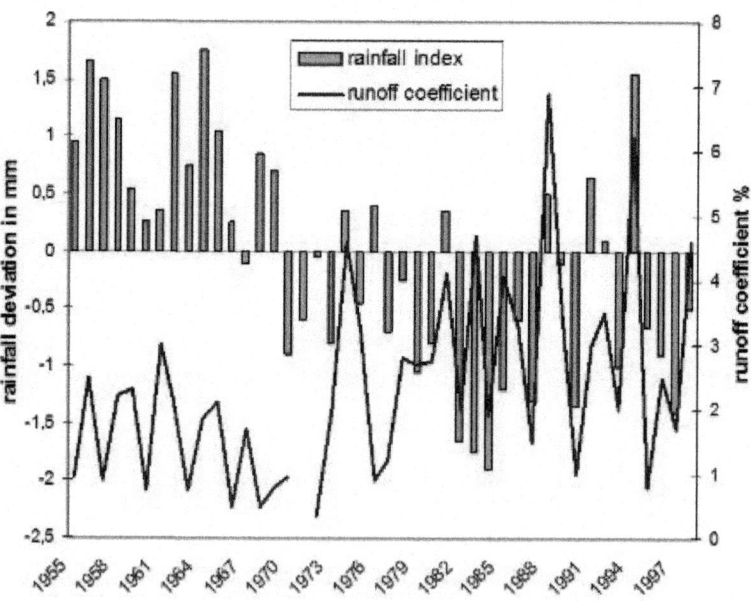

Figure 18: Runoff coefficients from observed measurements and rainfall indexes over the Nakambé river basin (Burkina Faso) [22].

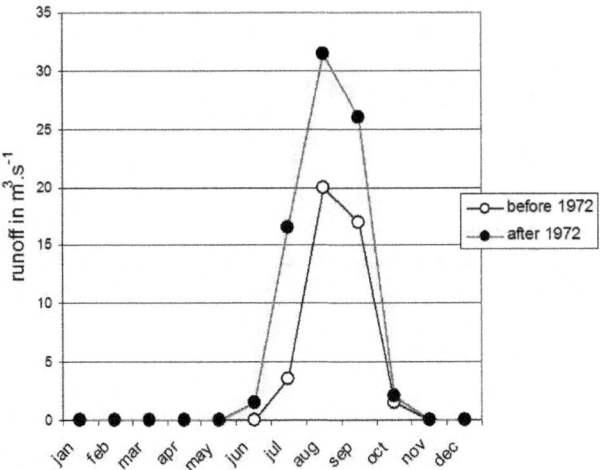

Figure 19: Monthly averaged discharges of seven Mauritanian tributaries of Senegal River, before and after 1972 [23].

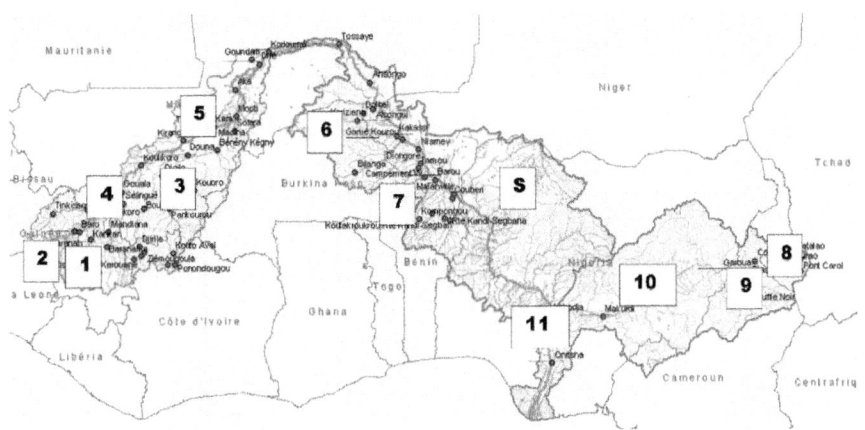

Figure 20: The Niger River basin [26]. Stations cited in Table 4 are underlined and the position of cited basins is indicated with its number in the table.

The synthesis presented in fig.20 and Table 4 shows that, except in Sahelian areas, the discharge has been decreasing as expected, since the beginning of the drought in 1968. Figure 21 shows that there is a clear regional distribution of hydrological behaviour, with the Sahelian region showing a strong increase in runoff, the Sudanian ones exhibiting a significant decrease in river discharge, while Guinean areas (rainfall > 1300 mm) generally show a slight decrease in discharges.

Table 4: mean discharges for two periods : before (until 1969) and after (since 1970) the Drought for 11 basins of Niger River [23]

Basin	Niger	Niger	Niger	Niger	Niger	Niger	Niger	Benue	Benue	Benue	Niger
River	Milo	Nian-dan	Bani	Niger	Niger	Sahelian basin	Mekrou	Mayo Kébi	Benue	Benue	Niger
Station	Kankan	Baro	Douna	Koulikoro	Diré	Rive droite	Barou	Cossi	Garoua	Makurdi	Onitsha
Country	Guinea	Guinea	Mali	Mali	Mali	Burkina-Niger	Niger	Cameroon	Cameroon	Nigeria	Nigeria
Area km²	9900	12600	101600	120000	366500	90500	10500	25100	64000	303600	1388300
Numéro fig1	1	2	3	4	5	6	7	8	9	10	11
years obs.	1950-2000	1950-2000	1922-2000	1907-2000	1924-2000	1956-1995	1961-1999	1955-2000	1946-1991	1955-1995	1950-1987
mean(-69)	211	271	639	1552	2244	35,4	41	104	388	3549	6651
mean(70-)	155	186	235	1039	1349	38,0	23	74	244	2816	5016
(70-)/(-69) %	-27	-31	-63	-33	-40	+7	-42	-30	-37	-21	-25

AND IN ENDORHEIC BASINS?

As the "Sahelian Paradox" seems to apply in the whole Sahel, it must apply also in endorheic areas. In certain endorheic areas in the Sahel, the water table level has been found to be rising over the last several decades despite the strong reduction in rainfall observed after 1968. This phenomenon has been previously defined as the "Niamey Paradox" [27]. The excess in runoff has significantly increased the number of ponds. While ponds are the main zones of deep infiltration, their increase explains the rise of the water table level (figure 22).

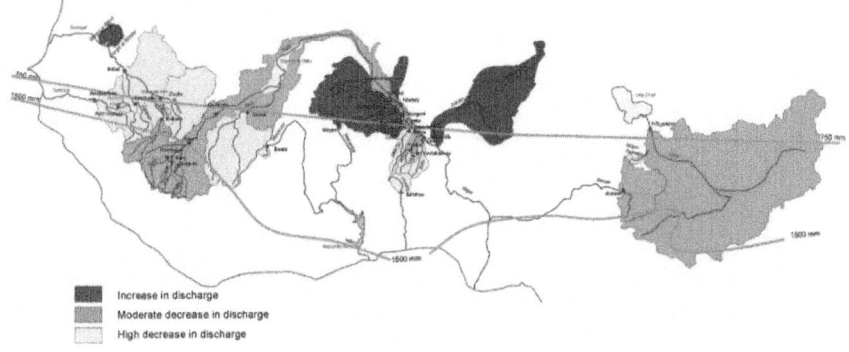

Increase in discharge
Moderate decrease in discharge
High decrease in discharge

Figure 21: The regional distribution of hydrological behaviour in West Africa; increasing discharges indicate a "hortonian" functioning as well expected decrease in discharge ([10] updated).

Indeed, the current active erosion processes are leading to the appearance of many new gullies and new spreading areas where sediments extracted by aeolian and hydric erosion are then transported in the gullies, and deposited. The gully beds are currently characterized by sand deposits ranging from 2 to 4 metres wide, several tens of centimetres deep (up to 1 to 2 m) and hundreds of metres long. Spreading areas are formed by these newly created streams when they reach gentler slopes, because their transport capacity becomes suddenly insufficient to carry such significant volumes of sand. They form sandy deposits of a magnitude of hundreds of square metres, up to several hectares, and, in some cases, tens of centimetres deep. These areas constitute new deep infiltration areas, accelerating the rise in water table [28].

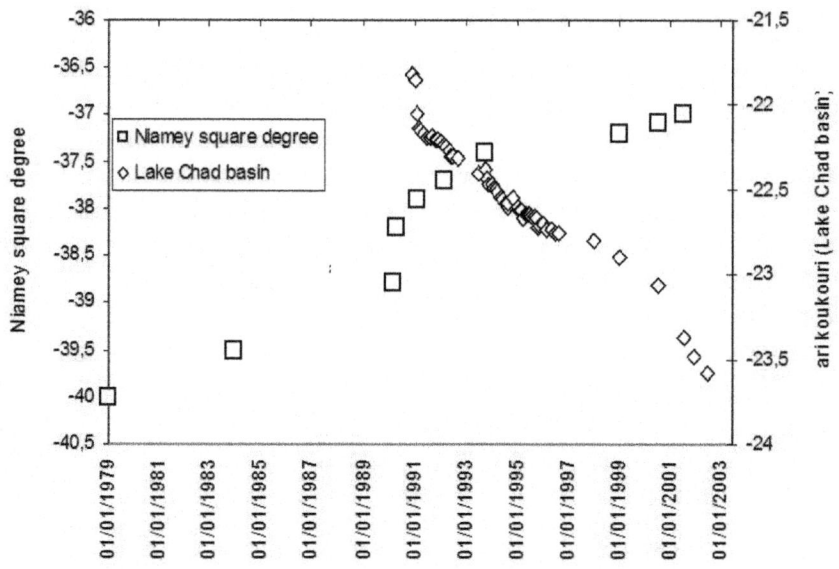

Figure 22: Evolution of water table level under the Niamey square degree area and under the Lake Chad [7] [28] [29], pers. Comm. of G. Favreau, 2007.

In contrast, the water table level is falling in the Lake Chad area (figure 22). This lake and the Niamey square degree are located at the same latitude, but the behaviours of their respective groundwater systems are completely different. The area of the Niamey square degree is composed of small endorheic basins with only local water contribution. While Lake Chad is mostly water fed by tropical humid areas of the Upper Logone and Chari basins (95%). Recently it was shown that discharges are decreasing in Sudanian areas [10] and, thus, water supply to Lake Chad is decreasing. The Ari Koukouri well is located at some tens kilometres of the northern part of the Lake, dried repetitively since the 1980s. The difference in the water feeding patterns explains the opposite behaviour of groundwater under Lake Chad and under the Niamey square degree.

A PERSISTENT DESERTIFICATION?

The evolution of land use/land cover in the Sahel remains the subject of an ongoing debate.

At the Sahel regional scale, cultivated areas have been increasing significantly for more than 50 years due to population growth and very low crop yields (see Fig. 23 for the Niger Republic) [30].Some studies based on satellite data vegetation indices [31] [32] [33] have suggested a re-greening of the Sahel after 1994. However, other studies ([34] amongst others) have highlighted the limitations of vegetation indices for the determination of land use and land use changes when using remote sensing at coarse resolution. Land cover studies based on aerial photograph analyses [28] show on the contrary a decrease in vegetation cover in south western Niger.

Although the Sahel is probably re-greening, the western part of Niger and the eastern part of Burkina Faso are still suffering land degradation (see maps in [32] [33]; see figure 24).

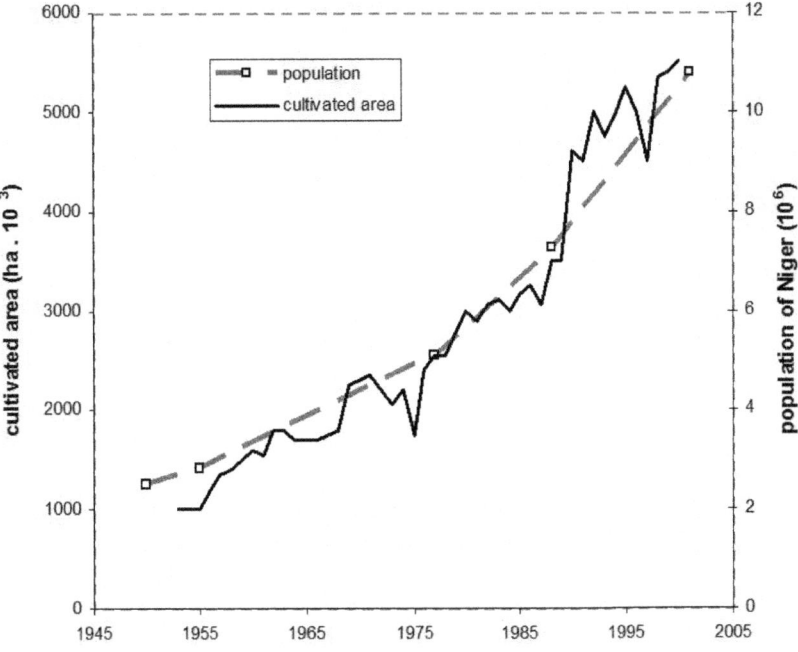

Figure 23: Evolution of population and cultivated area in Niger since 1950 [30].

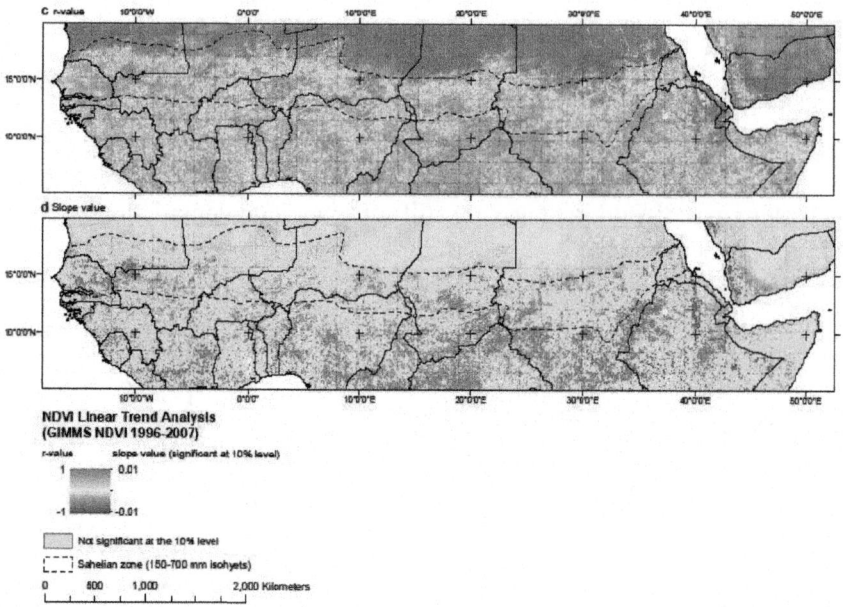

Figure 24: Evolution of NDVI values in the Sahel (area between the two doted lines) (up); down: mapping of the significant pixels (at 10% level) [33].

North of our study area, in eastern Mali, an observed increase in runoff has been attributed to the removal of vegetation during the drought and its inability to recover during recent wetter years due to soil erosion and degradation [35]. In this case, the drought is the cause of the new hydrological behaviour of the area.

Therefore, there is a series of problems to account for in order to improve our knowledge of the Sahelian evolution towards regreening or desertification:

1. The mapping methods to be used: before 1972, there was no satellite data. For this period only aerial photos can be used; therefore, there is no satellite product before the drought which begun in 1968. On the contrary nowadays, there is a large panel of satellite data available;

2. It is not always easy to segregate either the respective role of the drought (climatic, thus "natural" trend) and of the societies (demographic pressure on the resources) within the soil and vegetation degradation (when observed), or the impact of reclamation actions;

3. Studies based on aerial photos are very time-consuming and limited in space: thus, the selected areas for the analyses are not always representative of the entire region.

4.   The tools: the NDVI (normalised difference vegetation index) and the RUE (Rain Use Efficiency) are both indexes used to map the desertification trend.

5.   Some elements have to be accounted for such as the adaptation of vegetation to drought, the evolution of resilience in a drought context, the weight of graminaceae and herbaceous species in the measurement of NDVI and RUE, particularly in northern Sahelian areas where they represent 99% of the biomass, and then of the NDVI signal; in cultivated areas, the fact that pearl millet crops have a high biomass thus making the NDVI exceed that of the natural vegetation;

6.   Overall, there is a very high spatial and temporal variability in the evolution of land use/land cover, making a short - term or small – basin based comparative analysis difficult to generalise.

Finally, at the present time there is no certitude about the re-greening of the Sahel. However, hydrological studies showed that 5 to 10% of the land cover became crusted soils (ERO type, [14]) enough to significantly increase (twice) the runoff coefficient of a basin. Therefore, a general re-greening is consistent with the degradation of 5-10% of the total area, to explain the general increase in runoff in spite of an increase in vegetation cover.

CONCLUSION

A partial recovery of rainfall amount has been observed since the end of 20th century in the eastern and central parts of the Sahel. However, this cannot solely explain the increase in river discharges, since runoff coefficients have strongly increased (twice or more than those observed before the drought for basins of thousands of km²). Since the other rainfall characteristics did not show any change which could explain the rise in runoff and the earlier occurrence of the yearly flood, this is likely due to the land use change. Instead of deforestation, the land cover evolution is due to the increase in crop areas; since almost all the cultivable areas are now cultivated, the demographic pressure leads to a shortening of fallows. This is the main cause of soil degradation and crusting. Degraded areas are characterised by low infiltration capacity and high runoff coefficients.

The increase in discharge obviously is leading to a rise in flood hazard in Sahelian areas. This is also observed in Sudanian areas as during the monsoon of 2007 when thousands of km² where flooded in Burkina Faso, Togo and Ghana [32] [18]. However, this evolution is much more marked in the Sahel because runoff coefficients are rather decreasing in Sudanian areas since the beginning of the drought.

Furthermore, although crops have been destroyed in some areas of the Sahel, the flood hazard is becoming a severe planning problem in urban areas, where extended zones are tarred, leading to a strong reduction in infiltration. Urbanisation in flooding areas (figure 25) is creating a new land management problem. The Sahelian hydrological paradox thus has negative consequences rather than the sole positive opportunity of getting more water to agriculture and grazing. Crops commonly increase infiltration in the Sahel; extension of cropping on unsuitable soils, shortening of fallows and no-use of fertilizers are some of the causes of soil crusting and degradation.

Policy makers should be alerted to the effects of intensive cropping and land clearing, in some areas, on the hydrological regimes of Sahelian rivers. They must consider the flood hazard in downstream urban areas as an emerging severe concern to sub-Saharan populations.

Figure 25: Inundation of a recent housing area in northern Dakar suburbs (Senegal) (left); a street became a temporary river in Ziguinchor (Casamance, Southern Senegal, Sudanian area) (right); pictures taken during the 2012 monsoon.

ACKNOWLEDGEMENTS

We warmly acknowledge the colleagues of the Niger Basin Authority's "Projet Niger-Hycos" (NBA; http://nigerhycos.abn.ne/user-anon/htm/listStationByGroup.php) and the Niger Water Resources Department (DRE) for providing the Niger River discharge data. The Directors, and the colleagues of the Burkina Faso, the Mali and the Niger Meteorological Offices (Direction de la Météorologie Nationale) are also acknowledged. The SIEREM data were provided by A. l'Aour-Crès, N. Rouché and C. Dieulin (IRD-HSM), and the FRIEND animators provided rainfall data. This work was partly funded by French ANR projects ECLIS and ESCAPE, as well as by the AMMA project.

REFERENCES

1. Ali, A.; Lebel, T.. The Sahelian standardized rainfall index revisited. Int. J. Climatol. 2009, 29, 1705-1714.

2. Olivry, J-C. Synthèse des connaissances hydrologiques et potential en resources en eau du Fleuve Niger. Internal report World bank- Niger Basin Authority, 2002, 156 p.

3. Albergel, J. Sécheresse, désertification et ressources en eau de surface : application aux petits bassins du Burkina Faso. In "The Influence of Climate Change and Climatic Variability on the Hydrologic Regime and Water Resources"; IAHS publication N° 168, Wallingford, UK, 1987,pp. 355-365.

4. Mahe, G., Olivry, J.C. Assessment of freshwater yields to the ocean along the intertropical Atlantic coast of Africa". Comptes Rendus de l'Académie des Sciences, Series IIa 1999, 328, 621-626.

5. Amani, A.; Nguetora, M. Evidence d'une modification du régime hydrologique du fleuve Niger à Niamey. In "FRIEND 2002 Regional Hydrology: Bridging the Gap between Research and Practice", Proceedings of the Friend Conference, Cape Town, South Africa, 18-22 March, 2002; Van Lannen, H., Demuth, S., Eds.; IAHS publication N°274, Wallingford, UK, 2002, pp. 449-456.

6. Mahé, G.; Leduc, C.; Amani, A.; Paturel, J-E.; Girard, S.; Servat, E.; Dezetter, A. Augmentation récente du ruissellement de surface en région soudano sahélienne et impact sur les ressources en eau. In "Hydrology of the Mediterranean and Semi-Arid Regions, proceedings of an international symposium. Montpellier (France)", 2003/04/1-4, Servat E., Najem W., Leduc C., Shakeel A. (Ed.); Wallingford, UK, IAHS, 2003, publication n° 278, 2003, p. 215-222..

7. Descroix, L., Mahé, G., Lebel, T., G., Favreau, G., Galle, S., Gautier, E., Olivry, J-C., Albergel, J., Amogu, O., Cappelaere, B., Dessouassi, R., Diedhiou, A., Le Breton, E., Mamadou, I. Sighomnou, D. Spatio-Temporal Variability of Hydrological Regimes Around the Boundaries between Sahelian and Sudanian Areas of West Africa: A Synthesis. Journal of Hydrology, AMMA special issue, 2009,375, 90-102. doi: 10.1016/j.jhydrol.2008.12.012

8. Descroix, L., Laurent, J-P., Vauclin, M., Amogu, O., Boubkraoui, S., Ibrahim, B., Galle, S., Cappelaere, B., Bousquet, S., Mamadou, I., Le Breton, E., Lebel, T., Quantin, G., Ramier, D., Boulain, N. Experimental evidence of deep infiltration under sandy flats and gullies in the Sahel.

Journal of Hydrology, 2012, 424-425, 1-15;. http://dx.doi.org/10.1016/j. jhydrol.2011.11.019

9. Descroix, L., M. Esteves, K. Souley Yéro, J.-L. Rajot, M. Malam Abdou, S. Boubkraoui, J.-M. Lapetite, N. Dessay, I. Zin, O. Amogu, A. Bachir, I. Bouzou Moussa, E. Le Breton, and I. Mamadou. Runoff evolution according to land use change in a small Sahelian catchment. Hydrol. Earth Syst. Sci. Discuss., 2011, 8, 1569-1607, 2011www.hydrol-earth-syst-sci-discuss.net/8/1569/2011/doi:10.5194/hessd-8-1569-2011.

10. Amogu O., Descroix L., Yéro K.S., Le Breton E., Mamadou I., Ali A., Vischel T., Bader J.-C., Moussa I.B., Gautier E., Boubkraoui S., Belleudy P. Increasing River Flows in the Sahel?. Water, 2010, 2(2):170-199.

11. Descroix, L., Genthon, P., Amogu, O., Rajot, J-L., Sighomnou, D., Vauclin, M.. Change in Sahelian Rivers hydrograph: The case of recent red floods of the Niger River in the Niamey region. Global Planetary Change, 2012, 98-99, 18-30.

12. Hulme, M. Climatic perspectives on Sahelian desiccation : 1973–1998. Global Environmental Change 2001, 11, 19–29 http://dx.doi.org/10.1016/S0959-3780(00)00042-X.

13. Nicholson, S.E. On the question of the "recovery" of the rains in the West African Sahel. Journal of Arid Environments, 2005, 63, 615–641. doi:10.1016/j.jaridenv.2005.03.004

14. Casenave, A. & Valentin, C. A runoff capability classification system based on surface features criteria in semi-arid areas of West Africa. Journal of Hydrology, 1992, 130, 231–249 (1992).

15. Asseline, J., Noni, G.D., Chaume, R. Design and use of a low speed, remotely-controlled unmanned aerial vehicle (UAV) for remote sensing [French]. Photo Interprétation, 1999, 37, 3-9. http://www.documentation.ird.fr/hor/fdi:010026172.

16. Sighomnou, D. Evènements de crues du mois d'août 2012 sur le Niger. Projet GIRE 2, Autorité du bassin du Niger,, Niger Basin Authority, Niamey, 8 p. 2012.

17. Karambiri, H. Brève analyse fréquentielle de la pluie du 1er septembre 2009 à Ouagadougou (Burkina Faso). Note technique 2iE, 4 p. 2009.

18. Tschakert, P., Sagoe, R., Ofori-Darko, G. & Codjoe, S.M.. Floods in the Sahel: an analysis of anomalies, memory, and participatory learning. Climatic Change 2010, 103, 471-502. doi: 10.1007/s10584-009-9776-y.

19. Tarhule, A. Damaging rainfall and floodings: the other Sahel hazards. Climatic Change, 2005, 72, 355-377. doi: 10.1007/s10584-005-6792-4.

20. Di Baldassarre, G., A. Montanari, H. Lins, D. Koutsoyiannis, L. Brandimarte, and G. Blöschl. Flood fatalities in Africa: From diagnosis to mitigation, Geophys. Res. Lett., 2010, 37, L22402, doi:10.1029/2010GL045467.

21. Mamadou, I. la dynamique des koris et l'ensablement de leur lit et de celui du fleuve Niger dans la région de Niamey. PhD thesis, Paris 1 Panthéon Sorbonne University and Abdou Moumouni University of Niamey, 260 p. 2012.

22. Mahé, G.; Paturel, J.E.; Servat, E.; Conway, D.; Dezetter, A. Impact of land use change on soil water holding capacity and river modelling of the Nakambe River in Burkina-Faso. J. Hydrol. 2005, 300, 33-43.

23. Mahé, G., Lienou, G., Bamba, F., Paturel, J-E., Adeaga, O., Descroix, L., Mariko, A., Olivry, J-C., Sangaré, S., Ogilvie, A., Clanet, J-C. Le fleuve Niger et le changement climatique au cours des 100 dernières années. « Hydro-climatology variability and change (Proceedings of symposium held during IUGG 2011, Melbourne, Australia)"; IAHS pub. n° 344, 131-137. 2011.

24. Mahé, G.; Paturel, J-E. 1896-2006 Sahelian annual rainfall variability and runoff increase of Sahelian rivers. C.R. Geosciences, 2009, 341, 538-546.

25. Mahé, G. Surface/groundwater interactions in the Bani and Nakambe rivers, tributaries of the Niger and Volta basins. West Africa. Hydrol. Sci. J. 2009, 54, 704-712.

26. Boyer, J. F., Dieulin, C., Rouché, N., Crès, A. Servat, E. Paturel, J. E. & Mahé, G. SIEREM: an environmental information system for water resources. In: Climate Variability and Change – Hydrological Impacts (ed. by S. Demuth, A. Gustard, E. Planos, F. Scatena & E. Servat), 19–25. IAHS Publ. 308., 2006, IAHS Press, Wallingford, UK

27. Leduc, C., Favreau, G. and Shroeter, P. Long term rise in a Sahelian water-table: the Continental terminal in South West Niger. Journal of Hydrology 2001, 243, 43-54.

28. Leblanc, M., Favreau, G., Massuel, S., Tweed, S., Loireau, M., Cappelaere, B. Land clearance and hydrological change in the Sahel: SW Niger. Global and Planetary Change 2008, 61 (1-2), 49-62 (2008).

29. Zaïri, R. Etude géochimique et hydrodynamique de la nappe libre du Bassin du Lac Tchad dans les régions de Diffa (Niger oriental) et du Bornou (nord-est du Nigeria). PhD thesis, Montpellier 2 University, 210 p. 2008.

30.   Guengant, J.-P., Banoin, M. Dynamique des populations, disponibilités en terres et adaptation des régimes fonciers: le Niger. FAO-CICRED, Publ., Roma, Paris, 142 p., 2003. http://www.cicred.org/Eng/Publications/pdf/MonoNiger.pdf

31.   Anyamba, A. and Tucker, C.J. Analysis of Sahelian vegetation dynamics using NOAA-AVHRR NDVI data from 1981-2003. Journal of Arid Environments 2005, 63(3), 596-614.

32.   Prince, S.D., Wessels, K.J., Tucker, C.J., Nicholson, S.E. Desertification in the Sahel: a reinterpretation of a reinterpretation. Global Change Biology 2007,13, 1308–1313. doi:10.1111/j.1365-2486.2007.01356.x.

33.   Fensholt, R., Rasmussen, K. Analysis of trends in the Sahelian 'rain-use efficiency' using GIMMS NDVI, RFE and GPCP rainfall data. Remote Sensing of Environment 2011, 115, 438-451. doi:10.1016/j.rse.2010.09.014.

34.   Hein, L. & De Ritter, N. Desertification in the Sahel: a reinterpretation. Global Change Biology 2006, 12, 751-758.

35.   Gardelle, J., Hiernaux, P., Kergoat, L. & Grippa, M. Less rain, more water in ponds: a remote sensing study of the dynamics of surface waters from 1950 to present in pastoral Sahel. (Gourma region, Mali) Hydro. Earth Syst. Sci. 2010, 14, 309–324. doi:10.5194/hess-14-309-2010.

Chapter 11

CURRENT CHALLENGES IN EXPERIMENTAL WATERSHED HYDROLOGY

Wei-Zu Gu[1], Jiu-Fu Liu[1], Jia-Ju Lu[1] and Jay Frentress[2]

[1]Institute for Hydrology and Water Resources, Nanjing Hydraulic Research Institutes, Nanjing, China

[2]Oregon State University, Corvallis, USA

INTRODUCTION

The river basin, watershed or catchment is central to many of concepts in hydrology [1] including contaminant hydrology. In fact, two different studies in the Seine River basin during the end of seventeenth century, independently made by Pierre Perraut and Edme Mariotte have been identified by A. K. Biswas as the beginning of quantitative hydrology [2], Along with Perraut and Mariotte, Biswas also suggested Edmond Halley as co-founder of experimental hydrology. It was later accepted that scientific hydrology was founded on these two basin studies [3].

However, basin studies developed slowly until the end of nineteenth century when public demands accelerated. The first modern basin studies commenced in Emmental in Switzerland during the 1890s and were focused on hydrological differences between two small, ca 60 ha, basins [1]. A multitude of basin studies have appeared since the early twentieth century from many parts of the world. These include: (1) Wagon Wheel Gap experiment of USA begun in 1910 in two forested basins [4]; (2) Valday Hydrological Laboratory of USSR begun in 1933 and focused on field investigations of multiple hydrological parameters in watersheds with different scales [5]; (3) Coweeta Hydrologic Laboratory of USA set up since 1934 for forest hydrology and, ecological research in two main basins with 32 sub-watersheds under various treatments [6]; (4) Harz Mountains experiment of Germany with a pair of catchments, Wintertal and Large Bramke, begun in 1948 and focusing on hydrological effects of land use [7]; (5) Alrance experiment of France started in 1950 and focused on streamflow patterns[8]; and (6) Bluebrook Runoff Experiment of China established on 1953 and focused on drainage problems of the vast agricultural plain of the North Huai He River [9].

A period of rapid worldwide development of hydrological basin studies resulted from the International Hydrology Decade (IHD) Representative and Experimental Basin Programme 1965- 1974, with an estimated 3000 basin studies conducted during the Decade [1]. "Representative basins (RBs), which are selected as representative of a region of presumed hydrological similarity, are used for intensive investigations of specific problems of the hydrological cycle (or part thereof) under relatively stable, natural conditions" [10]. Experimental basins (EBs) are relatively homogeneous in soil and vegetation, have uniform physical characteristics, and are deliberately modified for study [10]. However, many EBs are just well-instrumented basins of small area [11]. Sometimes both RB and EB are included, as in the fruitful Plynlimon experiment in the UK [12].

If we designate the first phase of basin study (until ca the middle twentieth century) as foundational and the second phase (during/after the IHD) as developmental, it would appear that experimental watershed hydrology is inevitably going into a third phase of transition and innovation. Most field experiments in watershed science to date remain largely descriptive, with results that are difficult to generalize [13]. In effect, "… catchment hydrology is trapped in a dead-end track, a theoretical impasse" [14]. Experimental watershed studies, the core of watershed hydrology, are now confronting tremendous and even more complicated challenges, which are raised mainly from changing environment conditions and partly from the weakness of current watershed studies. This chapter provides an overview of the main challenges facing experimental watershed hydrology. Addressing these challenges will hopefully lead to substantial innovation in the field.

An important and instructive paper addressing a new vision of watershed hydrology [13] asked "What's wrong with the status quo?" If we posed this same question to the experimental hydrologist, the limitations of most existing RBs/EBs might be characterized as mainly twofold. First, study basin designs have been limited by the black box concept and many misconceptions (e.g., the linearity, non-heterogeneity, additivity of hydrologic systems etc.). Second, operation has been substantially bounded by the hydraulic conception of these watersheds as isolated hydrological systems. All of these watershed studies monitor only total runoff at the stream-outlet and the subsurface responses of the watershed are only estimated by hydrograph separation, etc. These characteristics undermine the formulation of a unified theory of watershed hydrology [14] and the development of watershed models [13,15].

There is a clear need to move beyond the status quo and expand from this narrow hydrological perspective to generate hypotheses governing general behavior across places and scales [13], with the ultimate aim "to advance the science of hydrology" [15]. For the third phase, the stage of transition and innovation, instead of classic RBs as described above, another kind of experimental watershed, the Critical Zone Experimental Block (CZEB) is suggested, with its concept and infrastructures very different from that of current EB. Following the concept of Critical Zone defined by the National Research Council [16], the CZEB will have practical boundaries for different geomorphologic regions.

An ongoing trial application of the CZEB approach is presented in this chapter.

CHALLENGES

Numerous challenges face researchers as they move forward to advance hydrology science using experimental watershed studies.

Demands of Advancing Hydrology Science

More than two decades ago, V. Klemes proclaimed that "hydrology is as yet lacking a solid scientific foundation needed for its development as a natural science" [17] and this sentiment has been reiterated by others. Bras and Eagleson said "in the modern science establishment, this niche is vacant" in their paper titled "Hydrology, the fogotten earth science" [18]. Jim Dooge asked "Is hydrology now an established science? Is hydrologic practice now firmly based on scientific principles?"[19]. McDonnell et al concluded that "the critical ideas and positive vision presented in that (Dooge) paper remain just as fresh, relevant and, unfortunately, very much unfulfilled" [13]. Sivapalan also suggested several steps toward a new unified hydrologic theory [14]. Substantial progress in hydrologic science "ultimately depends on new experimental work, new field observations, and new data collection networks" as summarized by Kirchner [15]. But, what shape will these new experiments and networks take?

Coupling processes. A watershed is an inherently dynamic ecological system composed of a variety of biotic and abiotic processes. For such systems, it is important not only to focus on hydrological process but also on linked ecological, biogeochemical, pedological and geological processes. Misunderstandings of the interrelatedness of these processes are commonplace (e.g., summarizing the net effect of all biological processes as evapotranspiration). The vegetation root system forms a dynamically complex net for preferential flowpaths. It changes runoff generation mechanisms not only for surface runoff but the subsurface flows, and the root system itself is effected via feedback from the various, changing flow patterns. Vegetation growth also adapts according to climate, infiltration, soil water and even evapotranspiration while feedback mechanisms reinforce the interconnectedness of the watershed system [20].

Innovative measurements. More than twenty years ago, JE Nash suggested that discovery in hydrology sciences has been limited by "a deficiency in our empiricism" and, "our tolerance of poor methods of observations" [21]. Since then, methods of observations have improved significantly. However, our observation techniques still inhibit hydrology and a fundamental change to systematic measurement programs is needed. The current observation programs are aimed mostly at classic watershed hydrological process. Instead we should focus on a holistic description of heterogeneity of not only the watershed hydrological process but all related processes as described above, as well as landscape properties and climate inputs.

Supporting and testing hydrological model. Physically based hydrological models are one of the most promising directions for advancing hydrologic science. Grayson et al concluded that "the models are enabling us to ask more questions, many of which are fundamental to our understanding of the natural systems that are the subject of our models" [22]. However, vital weaknesses in current watershed models exist. First, they are generally based on "well known small-scale theories such as Darcy's law and the Richards equation for coupled balance of mass and momentum" [13] and do not accurately capture the spatial heterogeneity, inherent non-linearity, and non-additive properties of natural watershed systems. Second, a large number of models are heavily over-parameterized, leading to equi-finality, wherein multiple combinations of parameters can yield equivalent results [13]. Grayson et al showed that comparable fits to a hydrograph could be achieved using a saturated overland flow model or Hortonian flow assumptions [22]. Kirchner argued that "parameter-rich models may succeed as mathematical marionettes dancing to match the calibration data even if their underlying premises are unrealistic" [15]. Kirchner concluded that if "the present trend away from physical processes and toward mathematistry ('blackboard hydrology') continues in hydrologic education and practice, hydrology will end up in a dead end as a science and become useless for applications" [17].

To achieve substantive improvements, models should be developed "in conjunction with, and as an integral part of, carefully planned and executed field studies established for the purpose of advancing our knowledge of the natural system"[22]. Key experimental basin requirements for model-generated hypothesis testing include: use of innovative basin designs and inclusion of sufficient variation in physiographic settings, hydroclimates, and watershed scales.

Hydrologic Replumbing and Natural Climate Oscillations

The National Research Council of the US identified anthropogenic perturbation and replumbing of the hydrologic cycle [23] as a fundamental challenge [16] in watershed science. Human influences dominate "the natural cycle of freshwater causing environmental changes that have been argued to move the planet to a new geologic era termed the 'Anthropocene'" [23,24].

Natural climate oscillations with observed large, abrupt events, the widespread millennial scale climate changes of the last glaciation (consisting of Dansgaard-Oeschger oscillations, Heinrich events etc.) are hypothesized to have been forced by North Atlantic atmosphere-ocean-ice interactions [25]. It has been argued that these oscillations are ultimately driven by variations in eccentricity, axial tilt and precession of the Earth's orbit as described in the Milankovitch theory. The decadal and multidecadal climate variations are generally linked with recognized dynamics, such as the El Niño Southern Oscillation (ENSO) [23].

Anthropogenic modification of the water cycle could "push the climate into new regimes" because "climate change has taken the climate system out of the repeated cycle of glacial-interglacial episodes" [23]. These alterations may accelerate the arrival of millennial scale events and trigger a "tipping point" transition resulting in "major climatic perturbations on time scales of decades to centuries"[26]. How might experimentalists, a small proportion of total hydrologists, address these challenges with progressive basin studies?

Long term monitoring. Well-established representative basins and benchmark experimental basins distributed across different physiographic and hydroclimatic conditions are critical for assessing hydrologic changes due to replumbing. Basins instrumented during or before the 1950's may also be used to determine patterns of rainfall redistribution, resultant stream flow and related biological issues.

Consequences of replumbed hydrology. Monitoring and unraveling the hydrological and ecological responses and the feedbacks due to the replumbed hydrological cycle from water conservancy projects, especially those in arid and semi-arid regions, is needed. Dams are perhaps the most dramatic examples of human capacity to transform nature in the name of development [27], but development programs and projects create both winners and losers [28]. Dams in arid and semi-arid basins, especially in endorheic basins, break the natural cycle, looting downstream groundwater and accelerating desertification [29]. Long distance water diversion may reasonably be viewed as an ecological "planning disaster"[30]. There are concerns that development is outpacing scientific understanding, a circumstance that may prove disastrous for the environment for the generations to come.

Fundamental studies. Various hydrological and biogeochemical fluxes through catchments are important to understanding the effects of hydrologic replumbing, but many of these are poorly understood and rarely measured. For example, various subsurface ecological and hydrological components cannot be measured directly and must instead be estimated from a small number of point measurements. Particular emphasis has previously been placed on the downward flux of catchment processes components (i.e., rainfall and discharge) rather than upward flux- land evaporation, transpiration, and upward recharge of groundwater flux, etc – even though "these fluxes serve as important regulators of the dynamics of the cycle"[23]. There are various unknown or poorly understood mechanisms regarding natural catchment behavior that are affected by the hydrologic replumbing occurring at local, regional and global scales. These include: (1) the old water paradox [31]; (2) network-like preferential flow[13], natural symmetry between water, landscape, soil and vegetation and the underlying organizing principles between soil, vegetation and other biotic elements [14]; (3) the combined mechanisms of the geological water-cycle and hydrologic water cycle in the shallow hydrosphere, as well as the deep circulation of groundwater, which go far beyond current hydrogeological boundaries [32]; (4) the puzzle of the missing sink for the remaining 40% of watershed carbon balance, which after years effort a consensus for it still has yet to emerge [33]; (5) the mechanisms involved in low flow hydrology and its relation to shallow and deep groundwater, regolith, land use, water quality and the biodiversity of aquatic ecosystems [34]; and (6) the coupling of basin geomorphologic processes with hydrologic and ecologic processes, including the poorly understood processes and mechanisms of environmental release, transport, and biological transformation of various contaminants in the unsaturated zone, shallow aquifers and deep aquifers.

Historical data on water replumbling. Historical data from sources other than hydrometric stage records will be very helpful to understand large-scale environmental changes, serving as a reference for the natural variation and anthropogenic impact. Figure 1 illustrates desertification in an endorheic basin as the result of human impacts. Four stages of man-earth relationship are illustrated, from the 'stage of natural harmony' until the stage of the environmental disasters, the punishment of nature (seeFigure 1 caption).

Design systematic experimental facilities. The dynamic systems currently facing experimental hydrologists exhibit a wide range of heterogeneity and process complexity across large spatiotemporal scales. There are multiple ways to overcome such scaling problems [13,14], but hydrometric measurement systems in natural basins still face challenges (e.g., the ability of current rain gauges with conventional deployment techniques to obtain precise precipitation inputs for natural basins remains controversial). It is necessary to have systematic experimental facilities consisting of laboratory physical models, hillslope "catchment", small-size natural watershed with uniformly dominant vegetation, small-size natural watershed with more complex vegetation, unchannelized catchments, nested sub-watersheds, and watersheds of large size. Such experimental facilities need to be situated in zones with different physiographic and hydroclimatic conditions. Obviously, this would require collaborative efforts at an international scale.

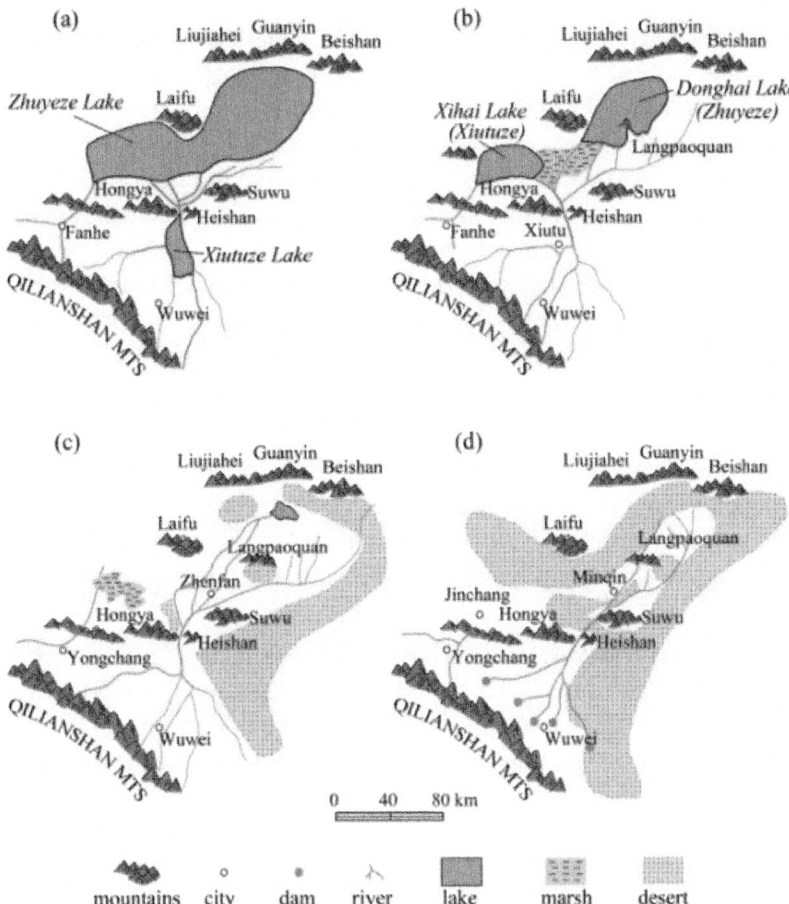

Figure 1: Desertification of the Rocksheep River (Shiyang River) Basin [35], an endorheic basin with total drainage area of 40,690 km^2. (a) prior to 121 BC, the "stage of natural harmony", the area of its terminal lake around 540 km^2; (b) 420- 589 AD, the "stage of relying"; (c) ca 1900AD, the "stage of impact", the cultivated area during Qing Dynasty was developed quickly, which reached to an area about 4/5 of that of modern time. (d) ca 1990's, the "stage of damage", there are 4 dams since 1959, 22 dams until 1972 [35].

THE CRITICAL ZONE EXPERIMENTAL BLOCK (CZEB)

For the future of hydrologic basin study, another kind of experimental basin is suggested: the Critical Zone Experimental *Block* (CZEB) customarily known as Critical Zone Experimental *Basin*.

Deficiencies in The Design and Operation of Current Experimental Basins

Watershed surface laterality. Ignoring the inherent interconnectedness of surface and subsurface watershed components reflects the fact that watershed study is driven by practical rather than scientific requirements. Hydrologists still lack a comprehensive understanding of surface responses to nonvisible and difficult to measure subsurface constraints. A physiological analogy might be studying the skin while ignoring the functions beneath. Focusing on surface responses using surface monitoring will quickly trap the observer with misconceptions. For instance, hydrologists delineate watersheds with surface topography, though actual watershed boundaries may cross topographically determined boundaries depending on subsurface features. Even our most fundamental measurements of the watershed – its shape and size – are strongly influenced by our inability to peer beneath the surface of the catchment.

Hydrologic process laterality. The hydrologic cycle is tightly coupled with other cycles [36]. The hydrologic cycle is also dynamic, not static, in nature. Studying the hydrologic cycle while ignoring the interconnectedness of other cycles is equivalent to studying blood circulation while only looking at blood vessels.

Downward components laterality. There is a tendency to emphasize measurements of downward components (e.g. rainfall, discharge and infiltration) rather than upward components (e.g. evaporation and transpiration aforementioned), even though upward components can be the dominant terms in water balance [37].

Lessons Learned From the Chinese Experimental Basin Studies Over The Last Fifty Years

The Chinese experimental watershed studies experienced a saddle-backed fluctuation with two peaks [9] during their fifty-year history. The lessons learned were summarized in cooperation with J.J. McDonnell, C. Kendall and N.E. Peters, as following [9]:

1.   The research facilities need not only the RB and EB of natural conditions, but also those with controlled boundary conditions. All EBs of recent

decades referred only to the natural boundary of the surface watershed, ignoring that of the bedrock. This frequently led to controversial results.

2. The EB should be treated as an integrated system, from surface to bedrock.

3. The runoff response of an EB, including surface and subsurface components, should be monitored hydrometrically. For all EBs in past decades, these components were monitored incompletely or incorrectly, insufficient even for 'black box' evaluations. As a consequence, most of the results can't be explained physically.

4. It is necessary to use multiple tracers to look inside the physical processes of the hydrological cycle and the anthropogenic impacts at the catchment scale.

5. Comparative EBs should be located in different climate and morphological regions and should have comparable monitoring strategies.

It was also suggested that hydrological experimentation should not address "only to the hydrosphere but the interactions between atmosphere, lithosphere, biosphere and the intelligence-sphere." To this end, there is a need to "develop approaches using physical, chemical, isotopic, biogeochemical and hydrometeorological methods for basin studies as a 'hybrid' basin research"[38].

What is CZEB?

The term "critical zone" has been used in many publications to refer to wide-ranging things in areas of geosciences and mineral, etc (e.g., the geological formation, the rhizosphere, the transitional zones in alluvial coastal plain rivers, etc) [36]. In 2001, the National Research Council of US recommended the integrated study of the "Critical Zone" (CZ) [36], defined as "the heterogeneous, near surface environment in which complex interactions involving rock, soil, water, air and living organisms regulate the natural habitat and determine availability of life sustaining resources" [16]. The CZ is one of the most compelling research areas in Earth sciences in the 21st century [36]. This zone was further defined as extending from "the vegetation canopy to the zone of groundwater" in 2005 [39], or "the bedrock to the atmosphere boundary layer" [40] and, "top of the vegetation down to the bottom of the aquifers" [36] by 2010.

The Critical Zone Experimental Block (CZEB) geologically is a monolith-block with its surface, the watershed, bounded by topographical water divides, which define the surface boundary, and which lacks defined subsurface boundaries. A drainage basin conceptually is only its surface part, the visible face (Figure 2). The CZEB is actually an Experimental "Block" within the Critical Zone with a surface drainage basin (the watershed). It is a dynamic ecosystem coupled with various supporting systems but using hydrological processes as the unifying theme. It is a living, breathing, evolving boundary layer where rock, soil, water, air and living organisms interact [41]. The CZEB is the experimental hydrological watershed study in the CZ framework. It follows that the components within the CZEB (including monolith, hillslope, sub-watershed, etc) will be monitored according to the «downward and upward approaches to theory development in catchment hydrology" [14, 42]. "The Critical Zone Observatory network is the only type to integrate biological and geological sciences so tightly"[40], and following this, the CZEB network will perhaps be the only network to integrate the hydrological, biological and geological sciences so tightly as well.

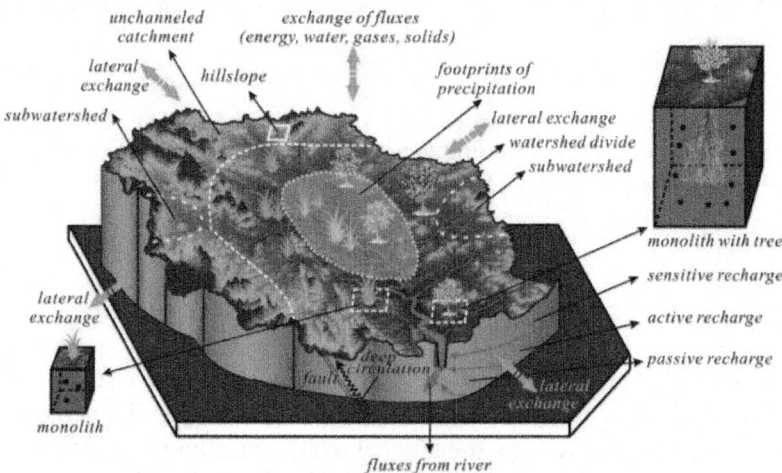

Figure 2: A conceptualized Critical Zone Experimental Block (CZEB).

Boundaries of Czeb

Top. The evaporation surface of the canopy in general is not the mean surface of the canopy but slightly lower. If the mean evaporation surface of the canopy is *h (m)* above the mean ground surface, then the top boundary of a CZEB is defined as *H(m)*, where *H*=(1.5 to 2.0)×*h* with coefficients of *h* varying according to the vegetation. This is mainly for the purpose of energy budget and eddy covariance flux observations. *H* is just the lower part of the atmosphere boundary layer.

Bottom. There are three cases:

Case I: Bedrock is situated at a relatively shallow depth from the ground surface while the regolith is shallow, too. The bottom boundary is defined as the geological boundary (Figure 3a).

Case II: Bedrock is deep. The bottom boundary is defined as the plane where the tritium content of groundwater approaches zero or the detection limit of ±0.7 TU (the "tritium naught line"(TNL)[43]), which is same as that of Case III (Figure 3b).

Case III: There are stratified alluvia, potentially with multiple aquifers with aquicludes and aquitards, common in flat areas with thick deposit and deep bedrock, up to hundreds of meters or more. Groundwater recharge to the river can be separated into sensitive, active and passive zones. In this case, the bottom boundary of CZEB is defined as the bottom of the active recharge zone, the plane of TNL (Figure 3c).

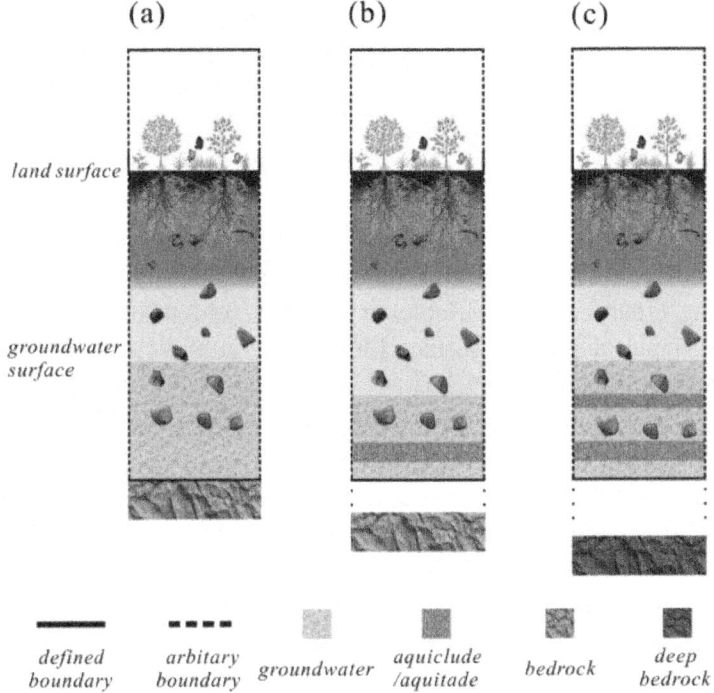

Figure 3: Boundaries of CZEB

Lateral sides

Part I: Above the ground surface up to height H as described above, delineated according to the surface topographic watershed boundary (Figure 3).

Part II: Below the ground surface, it is in general defined arbitrarily, except in the case of existing geological boundaries.

Functioning of Czeb

CZEB is an ecological dynamic and evolving system. It also is a natural open system, exchanging mass and energy with its surrounding environment across 'arbitrary' boundaries. It is a dissipative, complex system, however, with some degree of self-organization [36].

Interfaces. CZEB encompasses "the near-surface biosphere and atmosphere, the entire pedosphere, and the surface and near-surface portion of the hydrosphere and lithosphere" [36]. Within the CZEB, various processes including hydrologic, atmospheric, lithospheric, geomorphic and geochemical processes are coupled and dynamically interrelated. To simplify the organization of various interfacing processes throughout the CZEB, compartment zones can be broadly separated as follows:

1. The zone above ground surface, the "aboveground vegetation zone" [36].

2. The unsaturated zone, "the belowground root zone and the deeper vadose zone "[36].

3. The saturated zone, the "saturated aquifer zone" [36].

This layering has a general trend of increasing density with depth, has a dampening effect on state variables with depth, and an increase in distance to energy input at the soil surface [36]. There is also "an overall trend of increasing response time" [36, 44].

Mass – the material aspect of CZEB. The material base of the CZEB is fixed and limited, involving rock, soil, water, air and living organisms [16]. On this point it happens to coincide with the ancient Chinese philosophy, the so-called *"Five-xing"*, which holds that five fundamental elements form the universe and the Earth: "Jin"(metal), corresponding with the term "rock"; "Mu" (wood), with that of "living organisms"; "Shui"(water) with that of "water"; "Huo"(fire) with that of "air"; and "Tu"(soil).

Energy and force – the driven aspect of CZEB. Continuous energy fluxes and inherent forces drive the CZEB system. External solar energy is certainly the vital component of external energy source. Various inherent forces include: gravity, surface tension, intermolecular forces, capillary force, etc which control its dynamic situation.

Organization and entropy – the philosophical aspect of CZEB. This open system of structural dissipative processes and irreversible evolution tends to increase its entropy spontaneously and go towards disorder [36]. Even the exchange of entropy fluxes across its boundaries is continuous. The second law of thermodynamics, from which the concept of entropy was derived, is one of the fundamental natural laws. However, the role of feedbacks of this nonlinear system will promote "self-organization" as energy dissipates, providing opportunities for dissipating energy to act again within the system towards the direction of order. "Conservation without evolution is death. Evolution without conservation is madness" [45, 46]. Conservation of energy appears in the CZEB at all scales, with observable 'behaviors' of organization, symmetry, genes, etc. Lin suggested including "information" as one of the four general factors (i.e., conservation of energy and mass and, the accumulation of entropy and information, which dictates the evolutionary outcome and the functioning of the CZ) [36]. This is equally applicable to CZEB.

The Reactors of Czeb

The CZEB can be separated into three interdependent zones as mentioned above with each zone consisting of the same base materials, the *"five-xing"* rock, soil, water, air and living organisms as well as non-material bases of energy, force and information. This coincides well with the *'Bagua'* (*'eight-gua'*) of the earliest Chinese philosophical work *"I –Jing"* ("The Book of Chang"). According to I-Jing, the origin of the universe was just a singularity of chaos (*'Tai-ji'*), consisting of *'Yang'* and *'Yin'*, the unity of opposites, which produced *'eight-gua,'* the general principles governing development in the material world. Each *"gua"* is a combination of three whole lines (*'Yang'*) and broken lines (*'Yin'*). As in the four general characteristics of CZ elaborated by Lin [36]; in the CZEB, the material base refers to the five *'gua'*s (the *"five-xing"*) and the non-material base refers to another three *'gua'*s of energy, force and information, which together form a dynamic *'Bagua'*. Each zone has the same five material and three non-material components with different rates, Figure 4.

Each zone will behave as a 'feed-through reactor' according to Anderson et al [47]. It follows then that there are three feed-through reactors coupled together in the CZEB (Figure 4), with probably different rates and residence time of materials.

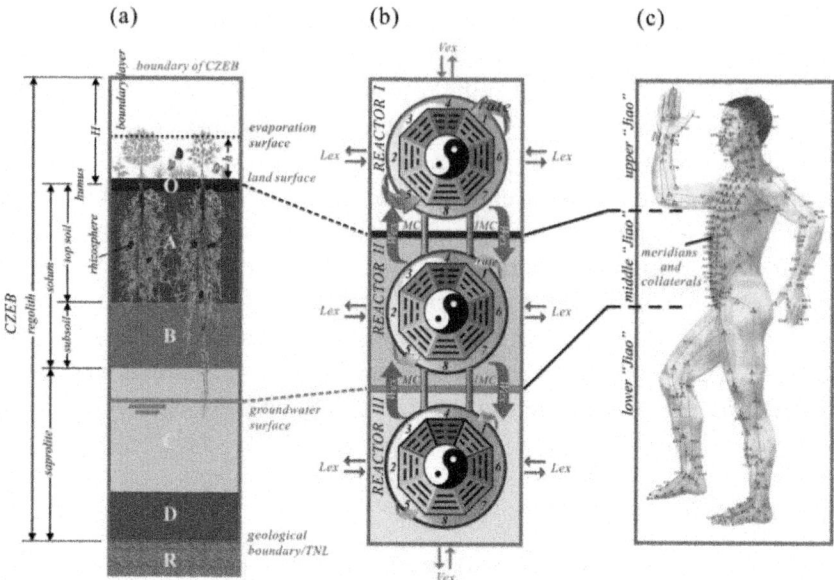

Figure 4: Schematic 'reactors' of the CZEB. The dynamic *'Bagua'* in each 'reactor' operates with its own rate; 1- 5 : the material base (1- air, 2- water, 3- soil, 4- rock, 5- organisms); 6-8: the immaterial base (6- energy, 7- force, 8- information).

Reactor I. The first zone is the above ground surface zone, which also contains above ground vegetation up to its interface with atmosphere (Figure 4a). Evaporation and plant transpiration may account for 50% or more of the total local precipitation and use up to 50% of the total solar energy.

Reactor II. The second zone is the unsaturated zone, extending from the soil surface to the upper surface of the groundwater table. It consists of the whole soil profile: The O horizon (humus), A horizon (topsoil), B horizon (subsoil) and C, and potentially D, horizons (Figure 4a). The unsaturated soil zone has been recognized as "the most complicated biomaterials on the planet" [36].

Reactor III. The third zone is the saturated zone, extending from the groundwater table down to bedrock, including the capillary fringe (Figure 4a). There are two general cases for CZEB, including phreatic groundwater and confined aquifers.

Functioning of reactors

1. Reactor I includes hydrometeorological and ecohydrological processes, Reactor II includes hydropedological and hydroecological processes, while the Reactor III is more exclusively for hydrogeological processes.

2. There is an overall trend of decreasing operation rates from Reactors I to III.

3. Each Reactor has its own lateral flux exchanges *Lex* via the CZEB lateral boundaries (Figure 4b). The current hydrometric runoff data is only that from channel. Reactor I has vertical fluxes exchange *Vex* with atmosphere via the upper boundary of CZEB while Reactor III has *Vex* with deep aquifers and/or deep circulation via the bottom boundary of CZEB (Figure 4b). Fluxes include all material and immaterial components.

4. Reactors are closely coupled within the boundaries of CZEB. There are flux exchanges *Wex*(Figure 5) between Reactor I and II and, and between II and III within the CZEB. There are flux channels *MC* (Figure 4) through I to III for materials, and flux channels *IMC* (Figure 4) through I to III for non-materials.

5. The *MC* appears similar to blood vessel system and *IMC* to meridians and collaterals of the human body from the Chinese art of acupuncture. The three "reactors" are similar to the so-called human "*three Jiaos*" (Figure 4), a central concept of Chinese medicine philosophy.

A TRIAL OF THE CZEB: THE CHUZHOU CZEB EXPERIMENTAL SYSTEM

The Strategy: A Two–Way Multi-Scale Approach

Isolation of the natural system. In order to investigate the natural watershed system, which is filled with complexity and heterogeneity, simplification is incorporated as much as possible (e.g., "*isolation*") [9, 38]. Using detailed observation on ever-decreasing scales allows the investigator to focus and isolate processes from the naturally occurring heterogeneity (i.e., watershed, sub-watershed, unchanneled watershed, hillslope, monolith, a tree) (Figure 5). To some extent it is similar to dissection: splitting "problems into their smallest possible components"[48].

There are limitations in the field, however. Using a single tree as an example, the tree itself can be well instrumented, as well as the subsurface with pits up to meters deep, but still the fluxes from its bottom boundary would be unknown. In this sense, *isolation* refers to a system rather than the physical isolation of a natural unit, which can only be partially isolated in field. Strictly speaking, the field result from such natural systems can be explained only empirically, from a statistic or a stochastic view. Understanding causality mechanisms from observational studies alone is problematic. Thus, McDonnell et al (2007) noted that "most field experiments and observations in watershed science to date remain largely descriptive" [13]. They also discussed a shared 'philosophical path' which claims that "if we characterize enough hillslopes and watersheds around the world through detailed experimentations, some new understanding is bound to emerge eventually", the authors concluded that "what this approach to experimental design has succeeded in doing is to help characterize the idiosyncrasies of more and more watersheds, in different places and at different scales, but with little progress toward realizing the Dooge vision" of "hydrologic laws"[13]. This does not mean, however, that field studies have no place in hydrology. On the contrary, the problem becomes how to "put the pieces back together again" [48]. In our case it is how to put together hydrologic mechanisms for a monolith, a slope, a sub-watershed and, a watershed.

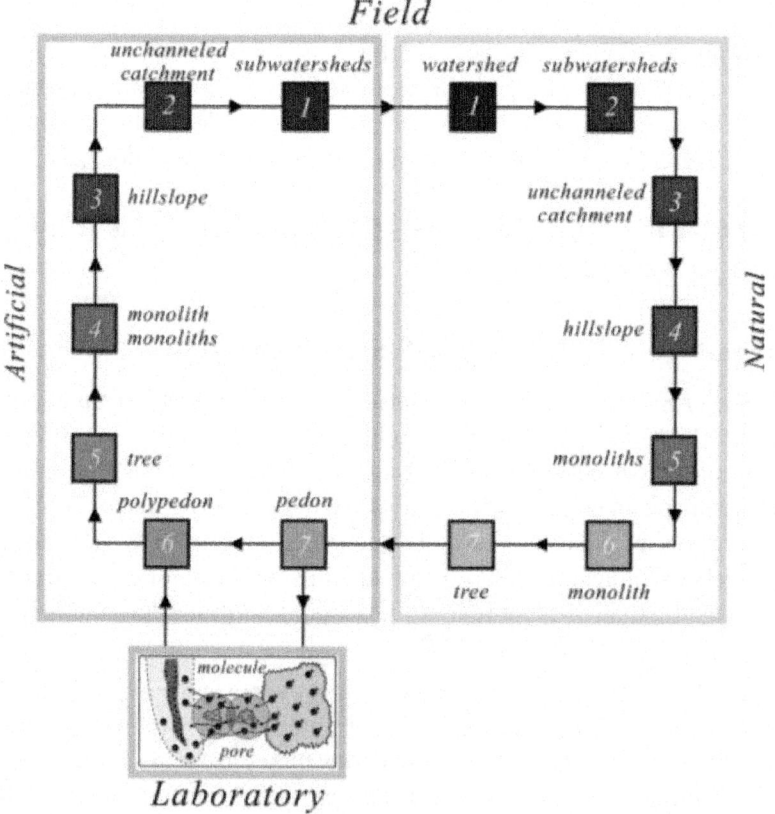

Figure 5: Concepts of the two-way multi-scale approach: natural and artificial.

Control and synthesis in the artificial system. A series of artificial systems are suggested and defined for "control and synthesis" [9, 38]. They encompass the pedon, polypedon, monolith, hillslope, catchment, until subwatershed

(Figure 5). Artificial systems are necessary in order to move beyond problems of boundary and parameter control found in natural systems. It is in fact inspired by the aforementioned empirical impasse of natural systems; artificial systems overcome heterogeneity by controlling all variables and leaving only test variables to fluctuate.

The laboratory system. The two-way multi-scale approach can only be applied at the macroscopic level of watersheds or, maybe the mesoscopic level (e.g., catena monolith). In fact many problems (e.g., the old water paradox) can be trace back to mechanisms occurring on a microscopic level (e.g., the molecular and pore related mechanisms as that of microbial processes of some contaminants transformation). The laboratory system connects with and supplements the artificial system (Figure 5).

Could this be a new vision for watershed hydrology and a unified theory of hydrology? Lin summarized three systems of different levels of complexity and a vast gap between the two extremes [36] as shown in Figure 6. The coupled natural and artificial systems are analogous to '*Yin*' and '*Yang*', which is the successful philosophy of '*Taiji*' (Figure 7). Klmes suggested "a rational search for meaningful conceptualization in hydrology can proceed along two routes: upwards and downwards "[42], or, the "upward or bottom-up approach" and the "downward or top-down approach" [14]. The two-way multi-scale approach perhaps could also be upwards and downwards. Hopefully, it could be a way for a new vision for watershed hydrology [13] and, for the unified theory of hydrology [14].

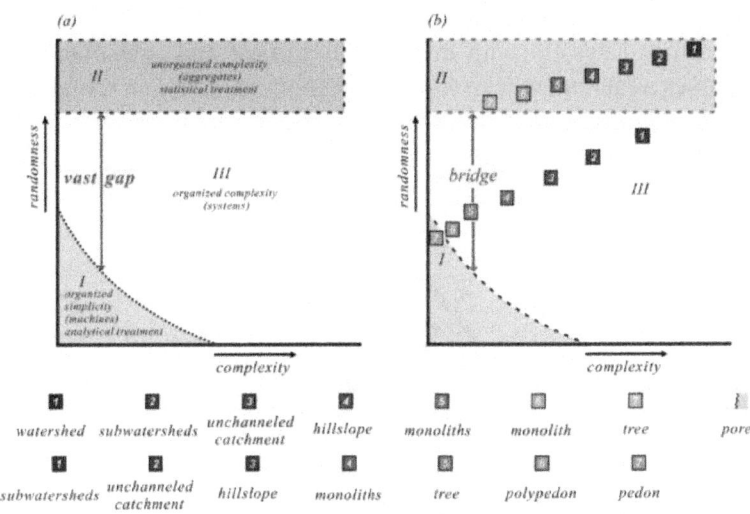

Figure 6: The gap between extremes of complex catchment systems. (a) Three types of complex catchment systems related to different treatment methods [14,19,36]; (b)

The supposed field and artificial two-way multi-scale framework and the gap depending on the level of control. The gap becomes the bridge.

The Trial Practice of The CZEB: The Chuzhou CZEB Experimental System

The Chuzhou CZEB can be attributed to the artistic conception of the *West Brook,* described in a poem of Tang Dynasty where intellectual exiles of Song Dynasty found their Arcadia ca 900 years ago. On the same brook in 1962, a hydrological experimental field station with three watersheds was founded. About fifteen years later, construction of the experimental watershed hydrology had barely survived from the cultural revolution of the 'Road to Perdition'. The Chuzhou CZEB was an adventurous plan for a two-way experimental watershed system including natural catchments and artificial catchments with controlled boundaries. The experiment was, unfortunately, interrupted again during the 'market-economy tides' of China and only partially completed. A turning point emerged a few years ago, with more efforts towards an innovative CZEB system. In the meantime, the components of the natural and artificial systems are: (1) the natural system with main infrastructure of hydrological monitoring and isotope biomonitoring encompasses watersheds with drainage area of 82.1 km² (hill 34%, forest 5%, ponds 1.2%), 17.5 km² (56%, 10%, 0.6% *idem*), 4.5 km² (100%, 2%, 0% *idem*), 3.3 km²(100%, 100%, 0%*idem*), 2.0 km² (5%, 2%, 30% *idem*), 0.06 km² (80% paddy field), 4573 m² (Morning Glory Catchment, 100% weathered debris), and, a mini CZEB with directly observable surface and subsurface responses with a drainage area of 7897m²; (2) the artificial system with measurable surface and subsurface responses encompasses a catchment with surficial and bottom area (horizontal projection) of 490m², a crop monolith of 150m², two grass monoliths (L1, L2) of 32 m² each, a saturated monolith (LS) of 1 m², and several weighing lysimeters. Here, the focus is on results from the two experimental watersheds - the mini CZEB watershed and the artificial catchment.

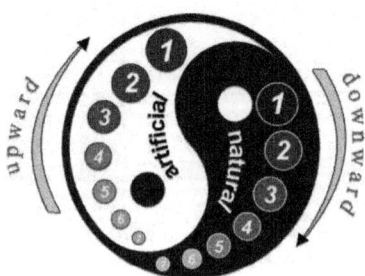

Figure 7: Reconciliation of two-way natural and artificial, *'Yin'* and *'Yang'* [14, 42].

The natural CZEB, Nandadish

Nandadish has a surficial drainage area of 7897 m² and rests on the consolidated bedrock of the concordant body of andesitic and tuffaceous facies with a thin weathered layer. The boundary condition of this mini CZEB belongs to the type shown in Figure 3a. The Quaternary regolith overlies the bedrock, its depth ranges from of 1 to 7 m with an average of 2.46 m; its bed rock topography was surveyed via 69 drillings (Figure 8). The regolith consists of brunisolic soil of heavy loam, medium and clay loams; saprolite with prismatic and block structures. Horizontal and vertical fissures and cracks developed in the upper regolith. This watershed has an altitude difference of 12.9 m with a surface slope of 6.7% to 17.1% at different directions. The brook gradient is 6.7%. The coverage during the watershed's construction in 1979 was natural grasses with small shrubs and a few Masson pines aged 5 to 6 years. Since then, coverage has shifted to a dense forest with canopy height ca 8 m.

The main infrastructure includes: (1) measurement for energy flux, water flux, geochemical flux, gas flux; (2) changes of watershed storages of energy, water, gas and that of geochemical ions; (3) CZ tree experiment; and (4) variations of precipitation isotope fingerprints in flux and in storage. The surface and subsurface runoff are monitored directly via a trench, which extends upslope to capture subsurface and surface flow. This CZEB block has deep soils near the divide but only 1-meter depths near the outlet, making the block easy to close via a concrete wall installed to the bedrock at the outlet point. In this way, all surface and subsurface flow drain into discharge measuring structures (Figure 9).

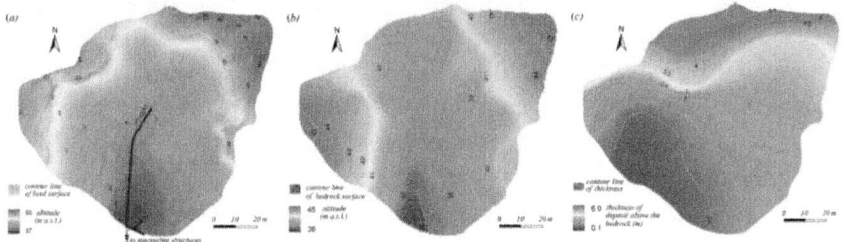

Figure 8: The mini CZEB Nandadish. (a) Its surficial watershed; (b) Its bedrock; (c) The depths of deposit above its bedrock.

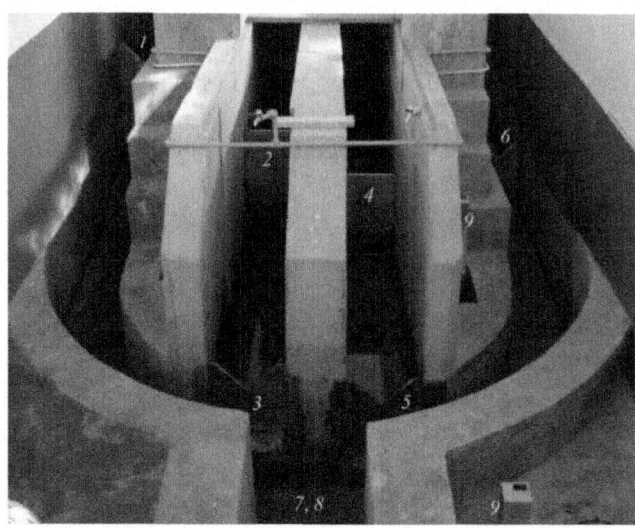

Figure 9: Discharge measurement structures for different runoff components from troughs with its location showing in Figure 8(a). 1- for rainfall; 2 and 3- for surface runoff; 4 and 5 – for interflow (30 cm below the soil surface); 6 – for interflow and groundwater flow (down to bedrock); 7 and 8 for total runoff (weirs are not shown). 1,3,5,6 and 8 are V-notch sharp-crested weir; 2, 4 and 7 are the full width rectangular sharp-crested weir; 9 is the tracking water head gauge. The picture design is the original concept, through renovations are now underway for gauge modernization.

The artificial catchment, Hydrohill

The Hydrohill catchment with a drainage area of 490 m² (horizontal projection by plane surveying) - 512 m² for its inclined surface [49, 50] - is sited on a small andesitic hill. The entire hillslope was first excavated to bedrock to create a bare catchment of ca 4700 m². A concrete aquiclude was created above it and consisted of two intersecting slopes dipping towards each other at 10°with an overall downslope gradients of 16.9°(Figure 10a). Impermeable concrete walls enclose the catchment on all sides to prevent any flow of water across the catchment boundaries. Silt-loam soil was removed, layer by layer, from an agricultural field near the Hydrohill site and installed on the artificial aquiclude, layer-by-layer again. Also the bulk density of the soil was adjusted during filling to approximate the natural soil profile of the original agricultural field. Hence, the final 1-m soil 'profile' was identical, at least in composition, to the natural soil profile. Grass was then planted over the surface and soil allowed to settle for 3 years (Figure 10b), after which time a central drainage trench was then excavated at the intersection of the two slopes and the water-sampling instrumentation was installed.

Five fiberglass troughs, each 40 cm wide and 40 m in length, were installed longitudinally in the trench. These troughs were stacked on top of each other to create a set of long zero-tension lysimeters (Figure 11a). Each trough has a 20 cm aluminium lip that extends horizontally into the soil layer to prevent leakage between layers. Water collected in troughs is routed through V-notch based logarithm weirs located in a gauging room (Figure 11b) under the hill where discharge is continuously monitored. Water samples are collected manually above the ponding at the weirs.

As illustrated in Figure 11a, the uppermost trough collects rain; the next lower trough collects surface runoff. The next three troughs collect subsurface flow from soil layers spanning the depths of 0-30, 30-60, and 60-100 cm. These troughs will be referred to as the 30 cm, 60 cm, and 100 cm troughs. The source of the water in these troughs (i.e., whether the water is derived from interflow or saturated flow) varies locally and during storms. The lowermost trough collects either saturated flow or interflow, depending on the height of the water table. When the water table is high, saturated flow may be collected in both of the lower two troughs.

A network of 21 aluminium alloy access tubes for neutron moisture gauges [51], a tensiometer scanner, and 22 wells for water-table measurements and groundwater sampling were installed (Figure 12). The wells were drilled to the aquiclude and are slotted along the lowermost 20 cm. After installation, the spaces around the slotted lengths were packed with sands to allow movement of groundwater to the well, space above the slotted lengths were packed with clay to prevent vertical drainage along the pipes (Figure 11c). The neutron probe access tubes (Figure 11c) and soil water potential tensiometers were positioned adjacent to the wells for soil water monitoring. Catchment evaporation was monitored by methods of water balance, energy budget and, soil water variations above zero flux potential. The plan view of the surface topography, the locations of wells and access tubes, and, the central stacked lysimeter troughs and that of energy budget set are showing in Figure 12.

Figure 10: The artificial catchment Hydrohill: (a) Its concrete aquiclude; (b) Three years waiting for the settling of filled soil to try to match close to the natural soil profile.

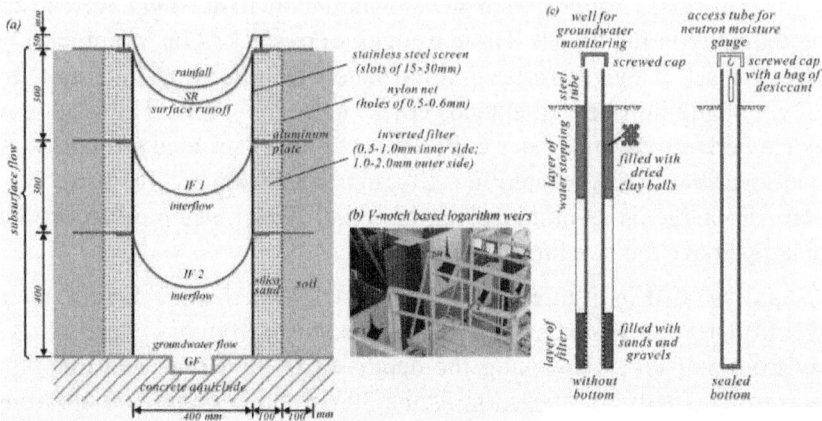

Figure 11: The artificial Hydrohill: (a) Schematic cross-section of rain, surface, and subsurface flow collectors at the catchment; (b) The V-notch based logarithm weirs for discharge measurement for these components; (c) The constructions of wells for groundwater monitoring and access tubes for neutron moisture gauge. (b) and (c) are the original design, now under renovation.

The renovation phase of the main infrastructures for this physically modeled CZEB includes: (1) measurement for energy flux, water flux, geochemical flux, gas flux; (2) changes of watershed storage of energy, water, gas and that of geochemical ions; (3) a small separate laboratory within the gauging room for high-frequency, real time measurement of isotope concentrations in all hydrologic components using of a laser spectrometer LWIA [52]; (4) a removable rainfall simulator on the tracks capable of covering an area of more than 500 m², capable of simulating a range of rainfall intensities; and (5) a system for carbon balance of this block.

SELECTED FINDINGS

Runoff Composition

The catchment response to precipitation consists of surface and subsurface components. These components were directly measured at Hydrohill with both isotopic and hydrochemical evidences (Figure 13) [53, 54]. Surface runoff was not always the dominant component during a rainfall-runoff event. The hydrograph composition of surface and subsurface flows was discriminated into four broad categories (Figure 14): 1) S type which is dominated by surface runoff; 2) SS type which is dominated by subsurface flows; 3) Intermediate type M, with largely equal contributions of surface and subsurface flows; and 4) Variation type V, with switching between components during an event.

Figure 12: Hydrometric monitoring of Hydrohill. (a) Plan view of the surface topograph showing the locations of groundwater wells, access tubes for neutron moisture probe with potential scanner, the central stacked lysimeter troughs, the energy budget set. (b) Whole view of its surface watershed in operation during early years (1980's), under renovation now.

Spatial and Temporal Variability in Hydrological Parameters

The physical based distributed-parameter modeling appears to be a promising direction for watershed hydrology. However the scale of model elements, and their parameters, is a lingering problem. This has led to investigations into the establishment of a 'representative elementary area' (REP), 1 km²[55]. In fact, because of the spatiotemporal heterogeneity of hydrological parameters, little is known regarding hydrologic parameterization in scales less than 1 km². Previously unimagined heterogeneities were observed at Hydrohill:

Intrastorm isotopic heterogeneity of shallow groundwater. During a storm in July 1989, the newly developed groundwater showed considerable spatial and temporal variability in the$\delta^{18}O$ (Figure 15a),with values ranging from -12‰ to -6‰ while the groundwater derived from pre-event soil water ranged from 0 to 100% at different times and locations in this catchment [56]. The groundwater samples for $\delta^{18}O$ were collected at three times during the storm from the wells distributed in whole area as shown in Figure 15a. At the end of the storm, the groundwater$\delta^{18}O$ values showed a 4‰ range in composition. This range of compositions during the storm appears to be caused by the combination effects of intrastorm variation in rain$\delta^{18}O$ (Figure 13) and spatial and temporal variability in subsurface flowpaths [56] – including the downward displacement of pre-storm water versus delivery of new water to the aquiclude via macropores [54].

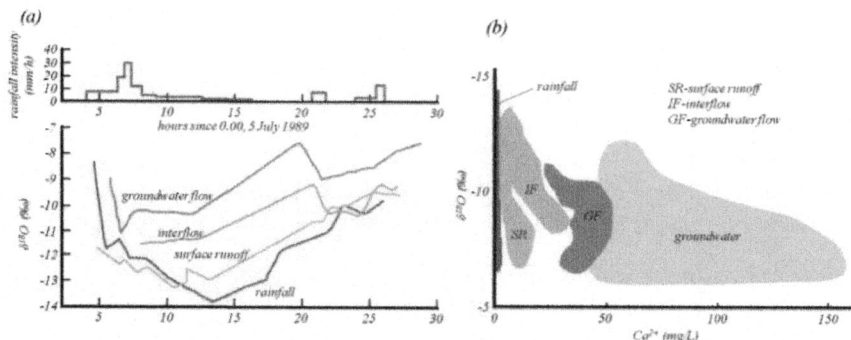

Figure 13: Isotopic and hydrochemical evidences of various runoff components. Samples for isotopes were analysed in Menlo Park laboratory of USGS by C. Kendall (same hereinafter).

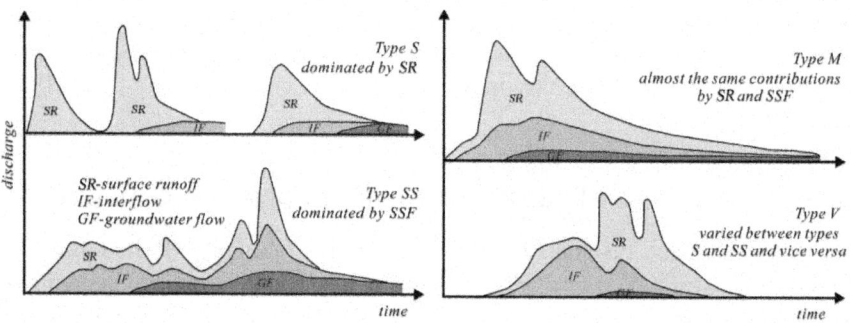

Figure 14: Types of runoff composition summarized from observed data of Hydrohill during 1981- 1992.

Annual heterogeneity of land evaporation rate Land evaporation was measured via the variations of unsaturated soil water at different positions in the soil profile during its matrix potential distribution in that profile showing a zero flux plane. Intensive measurements of volumetric soil moisture by neutron probe at 4 points within each of the 21 access tubes (84 in total) were made during the wet and dry seasons. Representative examples of heterogeneity in the daily evaporation rate are shown in Figure 15b [51].

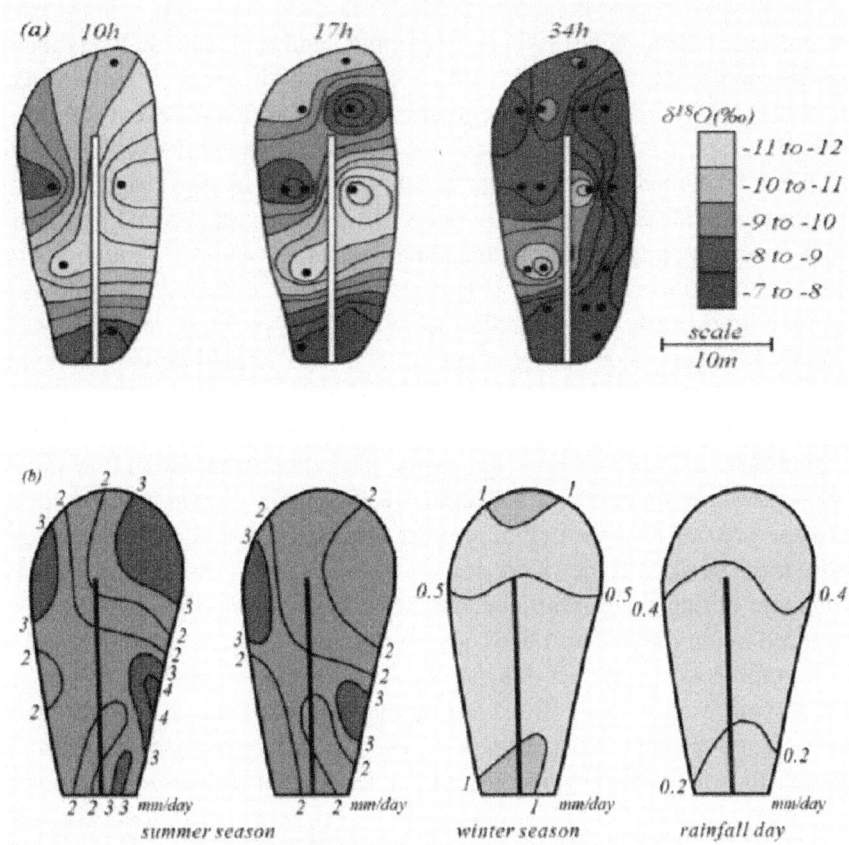

Figure 15: (a) Contour diagrams of $\delta^{18}O$ in groundwater at three sampling times and (b) Typical spatial distribution of the daily land evaporation rate.

The Double Paradox

About ten years ago, Kirchner described a double paradox in catchment hydrology and geochemistry [31]. His objective was to raise questions rather than provide answers and a worthy challenge was then put forward (i.e., "need to take up the search for a unified theory", so that the paradoxes "would no longer seem paradoxes") [31]. For this challenge, perhaps catchments with directly observable surface and subsurface components could be useful in two ways: First, to identify how this double paradox will emerge in the subsurface and surface components individually, instead of in the integrated form at the stream outlet; and secondly, to trace surface and subsurface generation mechanisms using isotopic fingerprints.

The delayed correspondency phenomena. Three isolated rainfall events are presented from May 1987 at Hydrohill, and are ideal for establishing rainfall-runoff relationships as well as the unit hydrograph (Figure 16a). In short, the unit hydrography conceptualization is that rainfall event A produces the hydrograph peak A, rainfall events B and C produce the peaks B and C etc. It can be named as the concept of *'one-to-one correspondency'*. However, using isotope tracers, rainfall-runoff events at Hydrohill showed that rainfall event A corresponded to the hydrography peaks A1 and A2, though A2 only emerged later during event B. This pattern continued throughout subsequent storms with event B corresponding to hydrograph peaks B1 and B2, though B2 only emerged later during event C. This was termed as the phenomena of *delayed correspondency* [57]. In the case of Hydrohill, this delay appears very distinct. It is presumed that such a *delayed correspondency* will tend to diminish as drainage area increases. The diminishment of the *delayed correspondency* phenomena has been verified in a natural watershed with drainage area of 82.1 km² (Figure 16b). Rainfall event A produces a runoff hydrograph peak A, some of the constituents of rainfall event A were delayed to emerge during the rainfall event B. It is worth noting that the hydrograph produced by rainfall event A does not have the bell shape referred to in the unit hydrograph concept. In fact this *delayed correspondency* shows the formation of pre-event water ("old" water) during the runoff process. It follows that the current *'one-to-one correspondency'*, used to conceptualize rainfall-runoff relationships in applied hydrology, will be associated with large uncertainty.

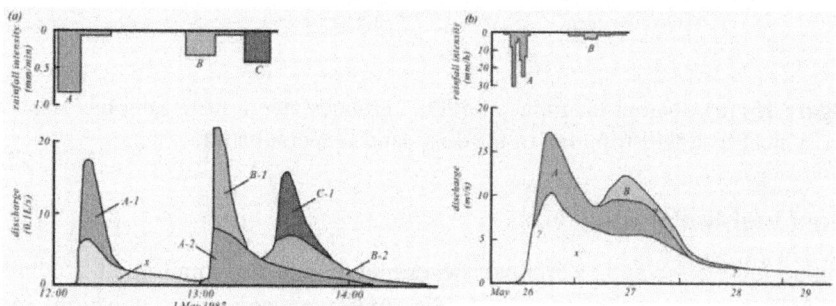

Figure 16: The delayed correspondency phenomena.

"Old" water paradox identified from both surface and subsurface flows. Data from the 'mini CZEB' watershed, Nandadish, suggest that pre-event water ("old water") appeared in all runoff components including surface runoff, interflow and groundwater or saturated flow (Figure 17). Figure 17a and Figure 17b refer to the surface runoff dominated type (type S, from above), and subsurface flow dominated type (type SS) respectively. The surface runoff

and subsurface runoff processes are shown separately with their corresponding proportions of pre-event water. For the type S, the pre-event water accounts for 9% and 24% of the total amount of surface runoff and of subsurface flow respectively while for the type SS, it becomes 11%and 89% respectively [58]. This reveals that even in a catchment with an average soil depth of 2.46 m, large volumes of pre-event water ("old" water) are stored and released promptly by event input.

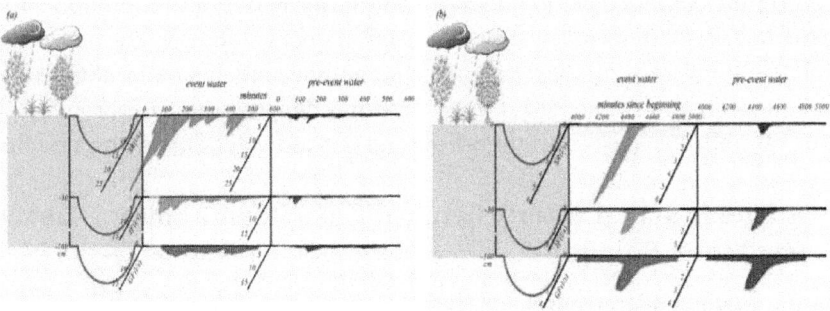

Figure 17: Processes of various runoff components, and that of their pre-event water of Nandadish. (a) for S type; (b) for SS type.

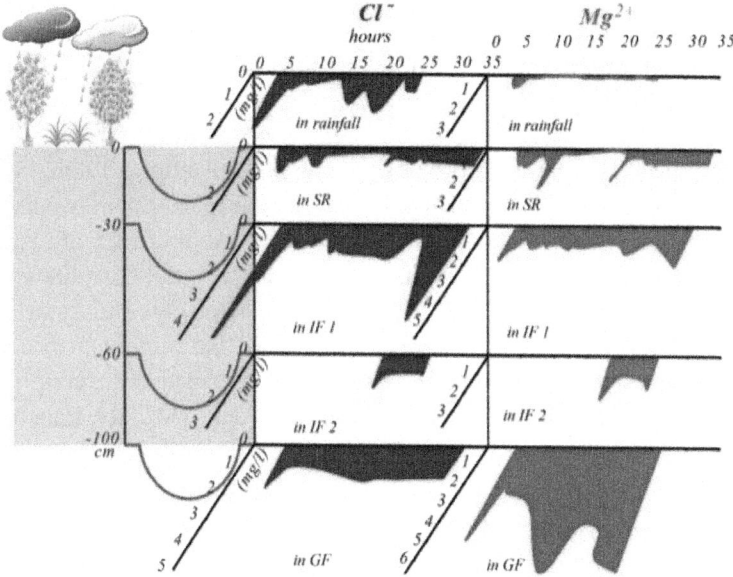

Figure 18: The Mg^{2+} and Cl^- processes in event rainfall and runoff components of Hydrohill. Samples were analysed in Atlanta laboratory of USGS by N.E. Peters (same hereinafter).

Hydrochemistry distribution paradox This was termed by Kirchner as the "'variable chemistry of old water' paradox: although baseflow and stormflow are both composed mostly of 'old' water, they often have very different chemical signatures" [31]. However, the artificial catchment at Hydrohill shows paradoxically the distribution of hydrochemical compositions in different runoff components, including event rainfall. This *'hydrochemistry distribution paradox'* results largely from: (1) inorganic ions in event rainfall input emerge in all runoff components; (2) a strong similarity between rainfall and surface runoff but less similarity in subsurface components; and (3) the fact that the total amount of ions of event rainfall is sometimes much smaller than the sum of all runoff components [59]. Figure 18shows the processes of Mg^{2+} and Cl^- in event rainfall, surface runoff, interflow and groundwater flow (saturated flow).

Does Diel Signal of Hydrochemical Constituents Emerge Linkage Among Multi–Processes?

For the better understanding of the links between contaminants and multi-processes, exploration of diel signals in natural waters may yield "insight into the intricate linkages among hydrological, biological, and geochemical processes [23]". Diel variations of various hydrological constituents in surface and subsurface runoff responses during rainfall events were monitored in artificial catchments and monoliths with examples as following (Figure 19).

Variations of pH. The diel variations of pH in SR, IF, and GF of Hydrohill show their own individuations (solid lines in Figure 19a). Diel variation curve of SR to a large extent appears similar to that of rainfall. However, pH variation curve of IF is contrary to that of rainfall after 10 a.m., while the GF curve appears as the flattened IF curve. The pH variation curve of the interflow of monolith L1 and that of the flow of monolith LS are reasonably similar to that of IF and GF of Hydrohill respectively (dot dash lines in Figure 19a). This reveals that the variation of pH in interflow and saturated flow is not driven only by rainfall input but also by "the biological processes of photosynthesis and respiration [22, 60]." This is ascribed to the coupling of the three "reactors" mentioned above via their MC and IMC for transformation and exchange (Figure 4).

Variation of ionic species. (1) For anion SO_4^{2-} : the diel variation curves of runoff responses of both Hydrohill and monoliths, with few exceptions, are contrary to that of rainfall. Mostly their peak concentrations happened during evening and midnight (Figure 19b). Different from the case of pH, SO_4^{2-} of SR has a strong inversion with the event rainfall curve at night time. (2) For cation Ca^{2+}: the diel curve of rainfall (Fig.19c) shows a small variation after sunrise. However, it triggers variations in runoff responses with their peak concentrations at both a.m. and before midnight. Highest peak happens to LI at afternoon. These diel variations in runoff perhaps are metabolism related. (3) Nimick et al [60] discussed diel cycles of dissolved trace metal concentration in a Rocky Mountain stream. They found that the anionic species "have their highest dissolved concentrations in the late afternoon" while the cationic species "have their highest dissolved concentrations shortly before sunrise"[23,60].

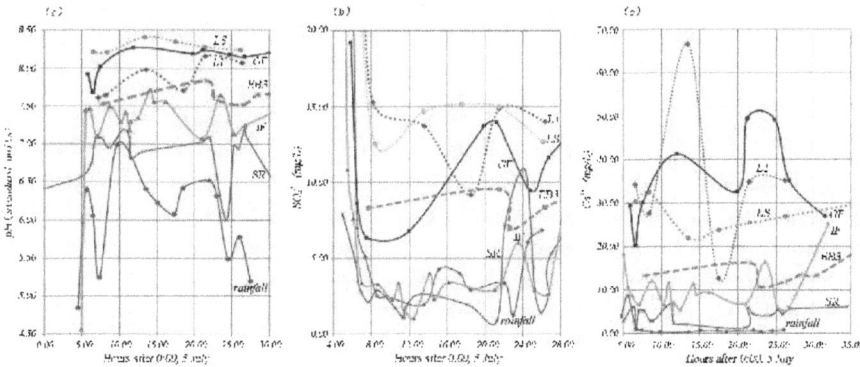

Figure 19: Diel variations of pH, and hydrochemical constituents SO_4^{2-}, Ca^{2+} in event rainfall and runoff responses of Hydrohill and monoliths. SR, IF, GF- runoff responses from Hydrohill; L1- monolith of 32 m^2 serves as an element slope of Hydrohill; LS - saturated monolith of 1m^2 serves as solum in reduction environment; EB3- Morning Glory Catchment of 4573 m^2 serves as a catchment without saturated zone formed by debris only.

Role of soil. Variations of pH, SO_4^{2-}, Ca^{2+} in total runoff of Morning Glory Catchment, a special designed catchment without soil but debris (EB3 in Figure 19 with broken lines), are very similar to that of rainfall curves after 20:00 (EB3 data are not enough before this time). The role of soil in the formation of ionic species in runoff is apparent. Even the variations of pH, SO_4^{2-}, Ca^{2+} in total runoff of EB3 are triggered by the event rainfall, but the resultant variations appear much simpler than various curves of Hydrohill and monoliths. This implies that only a simple process (i.e., mainly the hydrological process is involved in EB3). So, the complex diel variations of the runoff responses from Hydrohill and monoliths (Figure 19) can be reasoned as the results of multi-coupling among hydrological, biological, and geochemical processes. It shows that the reactor II (Figure 4) provides a key operator in contaminant hydrology.

Runoff Generation

The measurements of various runoff components, within artificial systems containing controlled boundaries, provide the possibility to look inside the formation mechanisms of individual components. Isotopic and geochemical tracing can help to investigate these mechanisms but only if significant differences in isotopic compositions of components occur. The general mechanisms for these runoff components (i.e., the surface runoff, interflow and groundwater flow (saturated flow)) are discussed in the following paragraphs [61,62].

Surface runoff (SR). Precipitation input is, of course, the essential condition for the generation of surface runoff. However, in order to actually generate runoff, there must be enough precipitation to form a thin, saturated soil layer (Lsat) at the ground surface. Saturation is key because unsaturated water movement is thought to be too slow to generate runoff. The thickness of Lsat at catchments, monoliths and plots was found to vary between 5 to 50 mm throughout events, increasing downward during rainfall events and receding once rainfall stopped, although these findings were complicated by the irregularities of the soil surface. SR was not generated until a Lsat was established at the surface. Once a Lsat was developed, regardless of its thickness, SR was generated immediately. Overland flowwas only observed on impermeable surface (DO) at artificial plot and, on saturated surface (SO) from a special lysimeter designed for Lsat simulation. Intrastorm variation of isotopic composition of DO andSO can match that of event rainfall. In most cases however, SR is generated within the Lsat, with turbulent mixing of event and pre-event water stored in Lsat, i.e., the saturated mixing surface flow(MS); Alternatively, small amounts of event water act on the surface of Lsat, to force out pre-event water in the Lsat, termed here as the saturated expelled surface flow (ES). The isotopic composition ofMS shows a mixing of event rainfall and pre-event water in Lsat. The isotopic composition of ES is similar to that of Lsat.

Interflow (IF) in the unsaturated zone. Three generation mechanisms are observed. (1) In cases there are soil layers with distinct bulk density and/or hydraulic conductivities, IF can be generated at the interface of soil layers. This was only observed in the 'mini CZEB', Nandadish, where IF occurred at the interface of the layers A (topsoil) and B (subsoil). This is termed as layered interflow (LI). (2) During percolation, soil water moves downward and laterally towards the drainage interface and accumulated until saturation. The saturation will expand if the soil matrix potential $\psi < 0$. Once ψ reaches zero, the accumulated saturated soil water will be released immediately at drainage interface. A small amount of infiltration water can trigger this process, to expel the accumulated soil water out as interflow. This process was observed for most IF events in Hydrohill and the 32 m2 monolith. This is termed as expelled interflow (EI). (3) Macropore interflow (MI) was also observed in a special designed catchment without saturated zone with area of 4573 m2.

Groundwater flow (GF) from saturated zone. In most cases, groundwater flow was due to a rising groundwater table resulted from event water recharge via the capillary fringe zone, and was termedrecharged groundwater flow (RG). RG was observed in most cases of both artificial and natural catchments. In addition, the macropore-induced groundwater flow (MG) is the only component not directly observed in this work but inferred from the isotope data in catchment GF during events.

Unreasonableness of Current Two-Component Isotope Hydrograph Separations

Hydrograph separation using conservative, two-component mixing models has been done in various natural basins. However, data from these two catchments, including both the natural and artificial systems indicate that this technique is unacceptable for natural basins and artificial catchments, yielding misleading results due to the unreasonableness of most of the assumptions involved and of the unrealistic operation procedures [63].

Violation of assumptions. In studying natural processes it is inevitable to make simplifying assumptions. However, as Kennedy et al [64] indicated, when simplifying assumptions are made, one runs the risk of drawing misleading conclusions. That may have been the case for the five assumptions [65] summarized for the use of isotopic techniques in precipitation-runoff studies.

The first assumption, that "Groundwater and baseflow are characterized by a single isotopic content" [65], is problematic. Baseflow in natural basins can be recharged by the active zone, passive zone, each with differing flowpaths for recharge resulting in differences in the isotopic composition of baseflow itself, which changes with time and discharge. It is also unreasonable to expect a single isotopic content for groundwater which is subject to both temporal and spatial variations (Figure 16).

The second assumption, that "Rain or snowmelt can be characterized by a single constant isotopic composition or, the variations are documented" [65], is problematic, because, as shown in Figure 14athe intra-storm variability of isotopic composition of rainfall is very significant. High isotopic variability has also been observed in experimental watersheds at Maimai, Panola, etc [56]. Additionally, the spatial variability of isotopic composition of rainfall in a watershed exists and shouldn't be ignored. Isotopic data from a 82.1 km^2 watershed showed that the largest difference in rainfallδ^{18}O reached 8.2‰.

The third assumption, that "The isotopic composition of rain water is significantly different from that of groundwater/baseflow" [65], can be true. However it must be demonstrated by appropriate sampling [64].

The fourth assumption, that "Contributions from soil water are negligible, or the isotopic composition is identical to that of groundwater" [65] is problematic, because it is a misconception. In a natural watershed, this can only occur if there is no unsaturated zone.

The fifth assumption, that "Contributions from surface water-bodies (such as ponds) are negligible" [65], is reasonable for most upland watersheds [64]. However, in catchments of hilly area with land use including series of ponds and paddy fields, the contributions are large enough to significantly influence results [62].

Based on such problematic assumptions, one of the basic conclusions that "most stormflow is old water"[66], which resulted from two-component hydrograph separations in multiple catchments with different drainage areas, appears suspect.

The physically ambiguous 'old' and 'new' water. This two-component separation model in fact is based on the classic Horton infiltration theory with a very simple runoff generation concept that the soil surface acts as a sieve capable of separating rainfall into two basic components [67]. This 'old' and 'new' water separation model has led to many usages with a variety of defined components in addition to 'old' and 'new' water component of flow, e.g., 'pre-event and event component', 'pre-storm and storm component', 'pre-storm water and rain in storm runoff', 'pre-event and rainfall water', 'groundwater component and event water', 'groundwater and rainwater', 'groundwater discharge and surface runoff'[68], 'surface and subsurface runoff'[69]. The classic two-component method has also been extrapolated to incorporate three components (e.g. that of channel precipitation, soil water and groundwater [70]). Thus, it seems that success can be assured because there are no calibration constraints. As Kirchner [15] noted some models are "often good mathematical marionettes, they can dance to the tune of the calibration data".

The 'old'(O) and 'new'(N) water actually correspond to different generation patterns of runoff components. In fact, multiple runoff mechanisms can result in the same proportions of old versus new water, a problem known as equi-finality. As seen in Hydrohill, surface and subsurface runoff patterns emerge from multiple runoff mechanisms (e.g., macropore flow can result in a large proportion of event water or a large proportion of pre-event water depending on antecedent wetness in the catchment). The runoff components corresponding to old water and new water can be very ambiguous. Applied to surface and subsurface runoff [68, 69], surface runoff is labeled new water [62]. This is a misconception as shown in Figure 20.

The mass balance equations for this model are untenable. The two-component tracer mixing model is stated by two mass balance equations for the composition of stormflow at any time which is used for hydrograph separation: $Q_s=Q_n+Q_o$ and $Q_s\delta_s=Q_n\delta_n+Q_o\delta_o$ where Q is streamflow, δ(Delta) is the D or ^{18}O content, the subscripts s, n and o represent the stream, new and old respectively[56]. At any time t, the first equation is correct (i.e., discharge $Q_{s(t)}$ at the outlet of a watershed is equal to the sum of the unknown $Q_{n(t)}$ and $Q_{o(t)}$ right at the same outlet at the time t). During time t, the $Q_{s(t)}$ at the outlet and its isotopic concentrationδ_s are measurable and are the value of known without problem, however both the $Q_{n(t)}$ and $Q_{o(t)}$ at the outlet of time t, are the results of confluence from somewhere within the watershed before time t, i.e., the results of a convolution integration. The $Q_{n(t)}$ and $Q_{o(t)}$ at the outlet of time t are concentrated from the separated isochrone area somewhere within the watershed with different time of concentration. So, the second equation with algebraic sum is physically unrealistic, hinging on the assumptions outlined above. For this linear algebraic sum to be physically realistic, the isotopic concentrations in the reservoirs of new and old water must be constant, without mixing or fractionating during the time interval of event hydrograph. To be physically realistic and representative of the mixing and non-conservative nature of water in natural catchments, it would likely take on the form of an integral or a finite-difference isochrone [63]. No affluxion happens to both the new and old water in the watershed. This can only happen in a small pond, operationally for a watershed, it's "the Procrustean bed [17]".

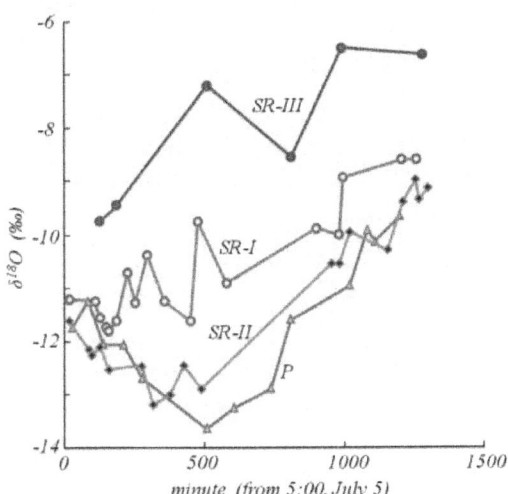

Figure 20: Intrastorm isotopic variations of an event rainfall, and isotopic variations of observed surface runoff in different watersheds. P–rainfall, SRI–surface runoff of the 'mini CZEB', Nandadish, with drainage area of 7897 m², SRII–surface runoff of Hydrohill of 490 m², SRIII- runoff of Morning Glory Catchment of 4573 m².

CONCLUSION

The river basin, watershed or catchment is a central concept in hydrology. Basin studies to assess watershed hydrology are approaching a period of transition and innovation. In fact, experimental watershed studies, the core of watershed hydrology, have tremendous and complicated challenges ahead.

The current challenges in basin experimental hydrology are mainly twofold. Advancing hydrologic science creates a fundamental challenge. Because the watershed system is a dynamic ecological system composed of a variety of biotic and abiotic processes driven by water and climatic processes, experimental watersheds should multi-couple these processes, organizing innovative measurements and approaches while continuing to support and test hydrological models. Anthropogenic hydrologic replumbing and natural climate oscillations are equally challenging. Field studies are the key to understanding and modeling the effect of hydrologic replumbing and climate change. There is a need for long term monitoring, systematic experimental facilities as well as data mining to reclaim historic data useful for determining baseline watershed metrics.

The main problems of the historical experimental basin approach are threefold: watershed surface laterality, hydrologic process laterality and downward components laterality. Lessons from fifty years of Chinese experimental basin studies are: (1) research facilities require natural conditions and also artificial controlled boundaries; (2) address not only the surface watershed, but also downward to the bedrock; (3) surface and subsurface runoff components should be directly monitored hydrometrically; (4) isotopic and hydrochemical tracers are key to understanding runoff generation mechanisms; and (5) account for interactions between the hydrosphere and multiple watershed processes.

Another kind of experimental basin is suggested going forward, namely the Critical Zone Experimental*Block* (CZEB), geologically a monolith-block within the Critical Zone. CZEB is a dynamic ecological and evolving system, coupled with various systems and united by hydrological process. The CZEB is a natural open system; both the energy and mass exchanges exist across its boundaries. It is a dissipative complex system with some degree of self-organization. The function of CZEB is threefold: mass, its material aspect; energy and force, its driven aspect and, organization and entropy, its thermodynamic/philosophical aspect.

To advance watershed hydrology and support development of a unified theory of hydrology, a two-way multi-scale experimental watershed system is suggested, including the natural system and the artificial system. Both have multi-scale subsystems from monolith, slope, sub-watershed to watershed and follow the research idea of upwards and downwards routes. A trial for such a strategy is partly completed at the Chuzhou CZEB Experimental System.

To advance contaminant hydrology, the suggested two-way multi-scale experimental watershed system may provide a key to unravel the complex mechanisms coupling hydrological, biological, and geochemical processes. Contaminant transformation and fate and their effects on regional degradation of groundwater basin involve highly complex mechanisms. This two-way multi-scale system calls for new models using both natural and artificial experimental basin results and using upwards and downwards approaches to multidisciplinary techniques, including isotope tracing.

All hydrological knowledge ultimately comes from observations, experiments, and measurements [15]. Progress in hydrology results mainly from challenges to prevailing approaches and concepts [71]. Hydrological experimentation, including CZEB experimental watershed studies, is the building block for development of a unified theory of hydrology including contaminant hydrology. However, it is important to remember Werner Heisenberg's warning that, "what we observe is not nature herself, but nature exposed

ACKNOWLEDGEMENTS

The renovation of the Chuzhou CZEB Experimental System is supported by the Hydrology Bureau of the Chinese Ministry of Water Resources and, the Nanjing Hydraulic Research Institutes. The adventurous plan for a two-way multi-scale hydrological experimental watershed system and its realization are led by Acadimician Jian-Yun Zhang.

Gu is deeply grateful to Jeffrey McDonnell and the group from USGS, Vance Kennedy, Carol Kendall, Norman (Jake) Peters for their kind help and support - the ever green cedars they planted by Hydrohill are now rooted and full of luxuriant foliage – as well as the wonderful water tracing methodology they taught, extending associations to new generation of colleagues and students. The authors are also deeply grateful to those who have visited Hydrohill for their teaching, help and support, they are: Sklash M. from Canada; Geyh M.A., Plate E. and Seiler K-P from Germany; Gat J. from Israel; Shiklomanov I.A. from Russia; Verhagen B. from South Africa, Littlewood I. from UK, Kinzelbach W. from Switzerland. Thanks to post-doctoral researchers Ma Tao, Xu J-T for their figures. Many thanks to the Editor of this book, Paul Bradley, for his kind help and, encouragements.

REFERENCES

1. J. C Rodda, Basin Studies. In: Rodda JC (ed.) Facets of Hydrology. London: John Wiley & Sons; 1976257297

2. A. K Biswas, History of Hydrology. North-Holland Publishing Company; 1970

3. Unesco/WMO/IAHSThree Centuries of Scientific Hydrology. Paris: Unesco; 1974

4. C. G Bates, A. J Henry, Forest and Stream-flow Experiment at Wagon Wheel Gap. Monthly Weather Review. Supplement.192830179

5. Урываев ВА. Зкспериментальные Гидрологические Исследования на Валдае. Ленинград: Гидрометеорологическое Иэдательство; 1953

6. J. E Douglass, M. D Hoover, Forest Hydrology and Ecology at Coweeta. New York: Springer-Verlag. 1988

7. J Delfs, W Friedrich, H Kiesekamp, A Wagenhoff, Der Einfluss des Waldes und des Kahlschlages auf den Abflussvorgane den Wasserhaushalt und den Bodenabtrag. Aus dem Walde. 19583223325

8. J Jacquet, Les Etudes d'Hydrologie Analytique sur Bassins Versant Experimentaux. Bull. Du Centre de Recherche et d'Essais de Chatou. 19622325

9. W-Z Gu, C-M Liu, X-F Song, J-J Yu, J Xia, Hydrological experimental system and environmental isotope tracing: a review on the occasion of the 50th Anniversary of Chinese basin studies and the 20th Anniversary of Chuzhou Hydrology Laboratory. In: Xi R-Z, Gu W-Z, Seiler K-P (eds.) Research basins and hydrological planning: proceedings of the International Conference on Research Basins and Hydrological Planning, 2231March 2004, Hefei, China. London: A.A. Balkema Publishers; 2004

10. C Toebes, V Ouryvaev, editors. Representative and Experimental Basins. Paris: Unesco; 1970

11. L. L Kelly, L. M Glymph, Experimental watersheds and hydrological research. In: Representative and Experimental Areas: proceedings of the International Symposium of Budapest, 28 September- 1 October. Paris: IASH; 1965

12. C Kirby, M. D Newson, K Gilman, editors. Plynlimon Research: The First Two Decades. Wallingford: Institute of Hydrology; 1991

13. J. J Mcdonnell, M Sivapalan, K Vache, S Dunn, G Grant, R Haggerty, C Hinz, R Hooper, J Kirchner, M. L Roderick, J Selker, M Weiler, Moving beyond Heterogeneity and Process Complexity: A new Vision for Watershed Hydrology. Water Resources Research 2007W07301

14. M Sivapalan, Pattern, Process and Function: Elements of a Unified Theory of Hydrology at the Catchment Scale. In: Anderson MG (ed.) Encyclopedia of Hydrological Science. John Wiley & Sons; 2005193219

15. J. W Kirchner, Getting the Right Answers for the Right Reasons: Linking Measurements, Analyses, and Models to Advance the Science of Hydrology. Water Resources Research 2006W03S04

16. NRCBasic Research Opportunities in Earth Science. Washington D.C.: The National Academies Press; 2001

17. V Klemes, Dilettantism in Hydrology: Transition or Destiny? Water Resources Research 1986S-188S

18. R Bras, P. S Eagleson, Hydrology, the Forgotten Earth Science. Eos 1987

19. J. C Dooge, Looking for Hydrologic Laws. Water Resources Research 1986S- 58S

20. P. S Eagleson, Climate, Soil, and Vegetation, 1. Introduction to Water Balance Dynamics. Water Resources Research 197814705712

21. J. E Nash, Foreward. Journal of Hydrology 1988v-viii.

22. R. B Grayson, I. D Moore, T. A Mcmahon, Physically Based Hydrologic Modeling. 2. Is the Concept Realistic? Water Resources Research 19922610026592666

23. National Research CouncilChallenges and Opportunities in the Hydrologic Sciences. Prepublication copy. Washington D.C.: The National Academies Press; 2012

24. G Vince, An Epoch Debate. Science 20113343237

25. R. B Alley, P. U Clark, L. D Keigwin, R. S Webb, Making Sense of Millennial-Scale Climate Change. In: Clark PU, Webb RS, Keigwin LD (eds.) Geophys. Monogr. Ser., 112Mechanisms of Global Climate Change at Millennial Time Scale. Washington D.C.: AGU; 1999

26. 26.   R. B Alley, P. A Mayewski, T Sowers, M Stuiver, K. C Taylor, P. U Clark, Holocene Climatic Instability: A Prominent, Widespread Event 8200 yr ago. Geology 199725483486

27. W. M Adams, Green Development, Environment and Sustainability in the Third World. London: Routledge; 2001

28. T Scudder, A Sociological Framework for the Analysis of New Land Settlements. In: Cernea M (ed.) Putting People First: Sociological Variables in Rural Development. Oxford: Oxford University Press; 1991148167

29. W-Z Gu, J-J Lu, The Disposition of Water Resources of Arid Area with Special Reference to the Alxa Plateau, Inner Mongolia. In: Maldini D,

Maher DM, Troppoli D, Studer M, Goebel J (eds.) Translating Scientific Results into Conservation Actions: New Roles, Challenges and Solutions for 21st Century Scientists. Boston: Earthwatch Institute; 2007

30. P Hall, Great Planning Disasters. London: Weidenfeld and Nicolson; 1980

31. J. W Kirchner, A Double Paradox in Catchment Hydrology and Geochemistry. Hydrological Processes 200317871874

32. W-Z Gu, Isotope Hydrology (in Chinese). Beijing: Science Press; 2011

33. W Steffen, P. J Crutzen, J. R Mcneill, The Anthropocene: Are Humans Now Overwhelming thr Great Forces of Nature? Ambio 200736614642

34. V. U Smakhtin, Low Flow Hydrology: A Review. Journal of Hydrology 2001

35. F-X Li, J-H Yao, Synthetical Research on the Economic Development and the Environmental Rectification for the He-Xi Corridor area (in Chinese). Beijing: Environmental Science Press; 1998

36. H Lin, Earth's Critical Zone and Hydropedology: Concepts, Characteristics, and Advances. Hydrology and Earth System Sciences 2010142545

37. M Falkenmark, J Rockstrom, Balancing Water for Humans and Nature: The New Approach in Ecohydrology. London: Earthscan; 2004

38. W-Z Gu, Experimental and representative basin stuides in China. In: Hooghart JC, Posthumus CWS, Warmerdam PMM. (eds.) Hydrological research basins and the environment: proceedings of the international conference, 2428September 1990, Wageningen, The Netherland. Hague: CHO; 1990

39. S. L Brantley, T. S White, A. F White, D Sparks, D Richter, K Pregitzer, L Derry, J Chorover, O Chadwick, R April, S Anderson, R Amundson, Frontiers in exploration of the Critical Zone: Report of a workshop sponsored by the National Science Foundation. 2005

40. Report prepared by the CZO CommunityFuture directions for Critical Zone observatory (CZO) science. 2010

41. National Critical Zone Observatory Program: http://criticalzoneorg/index.html

42. V Klemes, Conceptualization and Scale in Hydrology. Journal of Hydrology 1983

43. K-P Seiler, W Lindner, Near Surface and Deep Groundwater. Journal of Hydrology 19951653344

44. R. W Arnold, I Szabolcs, V. O Targulian, Global Soil Change. Laxenburg: International Institute for Applied Systems Analysis; 1990

45. G Bateson, Mind and Nature: A Necessary Unity. New York: Dutton; 1979

46. S Weinberg, Dreams of a Final Theory: The Scientist's Search for the Ultimate Laws of Nature. London: Vintage Books; 1994

47. S. P Anderson, F Von Blanckenburg, A. F White, Physical and Chemical Controls on the Critical Zone. Elements 20073315319

48. A Toffler, Forward: Science and Change. In: Prigogine I, Stengers I (eds.). Order out of Chaos. Toronto: Bantam Books; 1984

49. W-Z Gu, Field research on surface water and subsurface water relationships in an artificial experimental catchment. In: Dahlblom P, Lindh G. (eds.) Interaction between groundwater and surface water: proceedings of the international symposium, 30 May- 3 June, Ystad, Sweden. Lund: Lund University; 1988

50. W-Z Gu, J Freer, Patterns of surface and subsurface runoff generation. In: Leibundgut C (ed.) Tracer technologies for hydrological systems: proceedings of symposium H4 at the XXI General Assembly of the International Union of Geodesy and Geophysics, July 1995, Boulder, USA. IAHS Publ. 22919951995265273

51. W-Z Gu, Measurements of spatial evapotranspiration characteristics of an experimental basin using a neutron probe. In: Isotope techniques in water resources development 1987: proceedings of the International Symposium on the Use of Isotope Techniques in Water Resources Development, 30 March- 3 April, Vienna. Vienna: IAEA; 19871987789793

52. J Mcdonnell, Personal communications.

53. C Kendall, J. J Mcdonnell, W-Z Gu, A Look Inside 'Black Box' Hydrograph Separation Models: A Study at the Hydrohill Catchment. Hydrological Processes 20011518771902

54. C Kendall, W-Z Gu, Development of isotopically heterogeneous infiltration waters in an artificial catchment in Chuzhou, China. In: Isotope techniques in water resources development 1991: proceedings of the International Symposium on Isotope Techniques in Water Resources Development, 11-15 March, Vienna. Vienna: IAEA; 199219926173

55. E. F Wood, M Sivapalan, K. J Beven, L Band, Effects of Spatial Variability and Scale with Implications to Hydrologic Modeling. Journal of Hydrology 19881022947

56. C Kendall, J. J Mcdonnell, Effect of intrastorm isotopic heterogeneities of rainfall, soil water, and groundwater on runoff modeling. In: Tracers in hydrology: proceedings of the International Symposium on Tracers in Hydrology, 11-23 July 1993, Yokohama, Japan. IAHS Publ. 215199319934148

57. W-Z Gu, Experimental Research on Catchment Runoff Responses Traced by Environmental Isotopes. Advances in Water Science 1992in Chinese with English abstracts); 34246254

58. W-Z Gu, M-T Shang, S-Y Zhai, J-J Lu, F Jason, J. J Mcdonnell, C Kendall, Rainfall-runoff Paradox from A Natural Experimental Catchment. Advances in Water Science 2010in Chinese with English abstracts); 214471478

59. W-Z Gu, J-J Lu, X Zhao, N. E Peters, Responses of Hydrochemical Inorganic Ions in the Rainfall-runoff Processes of the Experimental Catchments and Its Significance for Tracing. Advances in Water Science 2007in Chinese with English abstracts); 18117

60. D. A Nimick, T. E Cleasby, R. B Mccleskey, Seasonality of Diel Cycles of Dissolved Trace Metal Concentrations in a Rocky Mountain Stream. Environmental Geology 2005475603614

61. W-Z Gu, Challenge on some rainfall-runoff conceptions traced by environmental isotopes in experimental catchments. In: Hotzl H, Werner A. (eds.) Tracer hydrology: proceedings of the 6th International Symposium on Water Tracing, 21-26 September 1992, Karlsruhe, Germany. Rotterdam: A.A. Balkema; 19921992397403

62. W-Z Gu, Various Patterns of Basin Runoff Generation Identified by Hydrological Experimentation and Water Tracing using Environmental Isotopes. Journal of Hydraulic Engineering 1995in Chinese with English abstracts); 5917

63. W-Z Gu, Unreasonableness of current two-component isotopic hydrograph separation for natural basins. In: Isotope techniques in water resources development 1995: proceedings of the International Symposium on Isotopes in Water Resources Development, 20-24 March, 1995, Vienna. Vienna: IAEA; 19961996261264

64. V. C Kennedy, C Kendall, G. W Zellweger, T. A Wyerman, R. J Avanzino, Determination of the Components of Stormflow using Water Chemistry and Environmental Isotopes, Mattole River Basin, California. Journal of Hydrology 198684107140

65. M. G Sklash, R. N Farvolden, The Use of Environmental Isotopes in the Study of High-runoff Episodes in Streams. In: Perry EC, Montgomery

CW. (eds.) Isotope Studies of Hydrologic Processes. Dekalb: N. Illinois University Press; 19826573

66. K. H Bishop, Episodic increases in stream acidity, catchment flow pathways and hydrograph separation. PhD thesis. University of Cambridge, Dep. Geol., Jesus College, Cambridge; 1991

67. R. J Chorley, The Hillslope Hydrological Cycle. In: Kirkby MJ. (ed.) Hillslope Hydrology. Chicheter: John Wiley & Sons; 1979142

68. E Crouzet, P Hubert, Oliver Ph, Siwertz E. Le Tritium dans les Mesures d'Hydrologie de Surface. Determination Experimentale du Coefficient de Ruissellement. Journal of Hydrology 197011217229

69. C Wels, R. J Cornett, B. D Lazerte, Hydrograph Separation: A Comparison of Geochemical and Isotope Tracers. Journal of Hydrology 1991122253274

70. D. R Dewalle, B. R Swistock, W. E Sharpe, Three Component Tracer Model for Stormflow on a Small Appalachian Forested Catchment. Journal of Hydrology 1988104301310

71. V Yevjevich, Misconceptions in Hydrology and Their Consequences. Water Resources Research 196842225232

Chapter 12

EVALUATION OF THE INFLUENCE CAUSED BY TUNNEL CONSTRUCTION ON GROUNDWATER ENVIRONMENT: A CASE STUDY OF TONGLUOSHAN TUNNEL, CHINA

Jian Liu, Dan Liu, and Kai Song

Faculty of Geosciences and Environmental Engineering, Southwest Jiaotong University, Chengdu, Sichuan 610031, China

ABSTRACT

Problems related to water inflow during tunnel construction are challenging to designers, workers, and management departments, as they can threaten tunneling project from safety, time, and economic aspects. Identifying the impacts on groundwater environment resulting from tunnel drainage and making a correct assessment before tunnel construction is essential to better understand troubles that would be encountered during tunnel excavation and helpful to adopt appropriate countermeasures to minimize the influences. This study presents an indicator system and quantifies each indicator of Tongluoshan tunnel, which is located in southwest China with a length of 5.2 km and mainly passes through carbonate rocks and sandstones, based on field investigation and related technological reports. Then, an evaluation is made using fuzzy comprehensive assessment method, with a result showing that it had influenced the local groundwater environment at a moderate degree. Information fed back from environmental investigation and hydrologic monitoring carried out during the main construction period proves the evaluation, as the flow of some springs and streams located beside the tunnel route was found experiencing an apparent decline.

INTRODUCTION

Nowadays, more and more tunnels are constructed for the reason that efficient transport strongly relies on road and railway tunnel, both in long-distance traffic and in metropolitan areas [1, 2]. However, when a tunnel interferes with groundwater in complex geological media, especially in carbonate karstic rocks, serious problems can arise during the excavation because of groundwater inflow, which is known as one of the most common but challenging problems faced by tunnel designers and constructors, leading to unsafe conditions, high construction costs, and delays, not to mention the risk to life and damage to property [1, 3–7]. Yuanliangshan tunnel on Chongqing-Huaihua railway in China came across three large filled caves at Maoba syncline carbonate strata, and a volume of approximately $4200\,m^3$ of clay erupted within 30 seconds in one cave of them, causing casualties and severe damage to the equipment in tunnel [8]. During the construction of Pinglin tunnel in Taiwan, the major difficulties encountered are caused by sudden high-pressure groundwater inflow, with a yield of approximately $180\,L/s$ in the pilot tunnel as an example, leading to the TBM being trapped and damaged and construction progress being greatly impacted [9]. Due to fluid drainage and pore pressure changes following tunnel construction, vertical settlements with magnitudes reaching $12\,cm$ were measured in fractured crystalline rock several hundred meters above the Gotthard highway tunnel in central Switzerland [10]. A series of hydrogeological problems with geotechnical and environmental impacts, causing spring discharges drying up and leading to a public protest, occurred during the construction of one of the high-speed railway tunnels between Malaga and Córdoba in south Spain [11]. Similar phenomena can also be observed during the drilling of the Firenzuola tunnel in Italy, water inrushes resulting in water table dropping below the level of the valleys and the gaining streams being transformed into losing streams or running completely dry, as did many springs, leading to severe damage to the aquatic fauna and other elements of the ecosystem [12].

Historically, groundwater studies associated with the design and construction of large underground structures have focused primarily on methods for control of groundwater inflows during excavation and for keeping the completed structure free of water. However, within the last several decades, the impacts of such activities on environment have become a major consideration. Though the impacts on groundwater resources of an area by underground excavation may be minimized by such planned constraints as preexcavation grouting and installation of an impermeable lining of the final excavation, such measures may not be efficient for avoiding claims of environmental impacts, particularly in areas of existing water shortages and/ or marginal supplies [13]. As any groundwater drawdown alters the natural hydrogeological flow system and can consequently impact groundwater-dependent vegetation, surface streams, lakes, wetlands, and associated aquatic ecosystems, as well as springs and wells, the wise approach is to assess the potential hydrological and ecological impact of a tunnel before building it and take appropriate measures to minimize the impact [12].

As mentioned above, groundwater inflow is one of the most complicated problems which can pose a serious risk and induce impacts on groundwater environment during the tunnel excavation. Accordingly, a number of measures must be adopted to minimize these effects. Appropriate measures can only be taken once the impacts are correctly identified, but it is a difficult task to obtain the correct identification [14]. Fortunately, some relative researches are trying to do this. For instance, a quantitative evaluation of hydrogeological impacts produced by a tunnel of 3 m in diameter and over 7 km in length in the surroundings of the city of Ferrol, NW of Spain, was assessed by means of calculating the hydrogeological behaviors before and after the tunnel excavation using water balance models, and then a comparison was made to allow for the quantitative evaluation of the changes in groundwater flow and the variation in the amount of water corresponding to each component of the model [14]; Tracer tests using uranine and sulforhodamine G and hydrological observations consisting of springs and streams were adopted to evaluate the effects of tunnel drainage on groundwater and surface waters in the Northern Apennines, Italy [12]; Yang et al. [15] utilized a numerical method and MODFLOW codes to simulate groundwater flow pattern in the tunnel area and determine the impact of tunneling excavation on hydrogeological environment in a regional area around the tunnel and a local hot springs area, at the "Tseng-Wen Reservoir Transbasin Diversion Project," in Taiwan.

Since the hydrologic and geologic system are very rare to be completely understood due to their complexity and heterogeneity, there are many difficulties and uncertainties existing in identifying the impacts caused by tunnel excavation on groundwater environment. Compared to problems, such as time delay and increase in costs, and losses to ecosystem and society, induced by drainage from tunnel, it is very essential to make an assessment of the negative effects caused by tunnel excavation prior to constructing it. Despite the fact that some efforts have been made to try achieving this goal, more works should be done to enrich the related researches.

This study presented in this paper focuses on evaluating the influence resulting from tunnel excavation on groundwater environment, by means of employing an indicator system proposed by Liu [16]. The procedure presented in this paper is applied to a case study of the Tongluoshan tunnel constructed in southwest China, where the karstification is well developed [17]. In order to completely understand the impacts caused by the excavation of Tongluoshan tunnel, some representative hydrological points and drainage from tunnel as well as precipitation in the tunnel area have been observed during most of the construction period.

METHODOLOGY

Indicator System for Assessment of the Influence Caused by Tunnel Construction

Water inflow is known as one of the most challenging problems during the tunnel construction, and many other hydrological and geological troubles such as regional water table drawdown, surface subsidence, and wells and springs drying up are induced by it. In order to assess the impacts caused by tunnel excavation on groundwater environment, indicators that closely related to water inflow into tunnel should be firstly taken into consideration. As both mining and tunneling are subsurface activities encountering the risk and damage produced by water inflow [1, 18], the same attention should be paid to the factors controlling water inrush to mine.

As concluded by Liu [16], the factors affecting water inflow during tunnel excavation can be classified into three categories including physical geography, geology and hydrogeology, and tunnel engineering. Each category can also be subdivided into several indicators which extremely explain how this category impacts tunnel inflow. In the category of physical geography, seven indicators such as average annual rainfall, average annual evaporation, area of catchment zone, coefficient of rainfall infiltration, relationship between the tunnel and geomorphology, capacities of reservoirs and lakes on the ground, and flow of surface rivers are included. It is necessary to note that catchment area does not always refer to the whole area of the hydrogeological unit that tunnel is located in but means a zone which collects water from precipitation and contributes to water inflow. Another source of water inflow is surface water, which should not be ignored because of its powerful ability to supply tunnel with abundant water. All the surface water contributing to water inflow should be taken into consideration when quantifying the indicators including capacities of reservoirs and lakes on the ground and flow of surface rivers. The category of geology and hydrogeology is composed of seven indicators too, which are carbonate rocks exposure ratio, water yield properties of the aquifers, water pressure on the tunnel, development of folds, development of fracture zones, formation lithology, and location of tunnel in horizontal and vertical hydrodynamic zoning of groundwater. The last category, tunnel engineering, consists of length of tunnel, area of disturbed range, construction method, burial depth of tunnel, and measures for prevention of groundwater flowing into tunnel. It is worth noting that water inflow does not always increase or decrease over burial depth of tunnel, but a special range of depths may be suitable for groundwater flowing toward tunnel.

The structure of the indicator system constructed by Liu [16] aiming at assessing the negative effects caused by tunnel excavation can be seen in Table 1.

Table 1: Indicator system for assessment of the negative effects caused by tunnel excavation on groundwater environment [16].

Objective layer	Rule layer	Indicator layer	Definition and explanation
[A] assessment of the negative effects caused by tunnel excavation on groundwater environment	[B₁] physical geography	[C₁₁] average annual rainfall (mm)	Average value of annual precipitation in the previous five to ten years
		[C₁₂] average annual evaporation (mm)	Average value of annual evaporation in the previous five to ten years
		[C₁₃] area of catchment zone (km²)	Area of the catchment zone that collects water contributing to water inrush into tunnel
		[C₁₄] coefficient of rainfall infiltration	The proportion of atmospheric precipitation contributing to groundwater recharge
		[C₁₅] spatial relationship between the tunnel and geomorphology	Spatial relationships between tunnel and geomorphology on cross and longitudinal section
		[C₁₆] capacities of reservoirs and lakes on the ground (m³)	Capacities of reservoirs and lakes, located on the ground, which may become water sources of tunnel inflow
		[C₁₇] flow of surface rivers (m³/s)	Flow of surface rivers which may supply tunnel with water
	[B₂] geology and hydrogeology	[C₂₁] carbonate rocks exposure ratio (%)	Areal ratio of outcropping carbonate rocks to the catchment zone contributing to water inflow in the plane
		[C₂₂] water yield property of aquifers	Water yield property of aquifers that may provide tunnel with water
		[C₂₃] water pressure on the tunnel (Mpa)	Hydrostatic pressure on tunnel
		[C₂₄] development of folds	Characteristics and scale of folds, as well as the development of water passages formed during folds formation
		[C₂₅] development of fracture zones	Development of fracture zones which may become water channels primarily including faults-fracture zone, joints concentrated zone, and contact zone of different lithology
		[C₂₆] formation lithology	Strata lithologic and its proportion
		[C₂₇] location of tunnel in horizontal and vertical hydrodynamic zoning of groundwater	Location of tunnel in horizontal hydrodynamic zone of groundwater including recharge zone, runoff zone and discharge zone, and in vertical hydrodynamic zone of groundwater, consisting of epikarst zone, aeration zone, seasonal fluctuation zone, shallow saturation zone, stressful saturation zone and deep circulation zone
	[B₃] tunnel engineers	[C₃₁] length of tunnel (km)	Length along the tunnel axis
		[C₃₂] area of disturbed range (m²)	Area of the zone that may be disturbed by tunnel excavation
		[C₃₃] construction method	Methods used to excavate tunnel, mainly including drilling and blasting method, New Austrian Tunneling Method, and tunnel boring machine method
		[C₃₄] burial depth of tunnel (m)	Vertical distance from ceiling of the tunnel to ground surface
		[C₃₅] measures for prevention of groundwater flowing into tunnel	Ideas and technologies adopted to prevent and treat groundwater flowing into tunnel

Fuzzy Comprehensive Assessment

The concept of fuzzy sets describing imprecision or vagueness was introduced by Zadeh [19]. Fuzzy logic where an element can belong partially to several subsets may be regarded as an extension of classical Boolean logic where belonging or not to a set is mutually exclusive. It simplifies the process of taking decisions by simulating the way of reasoning of a human expert in environments characterized by uncertainty and imprecision. Fuzzy evaluation methods process all the components according to predetermined weights and decrease the fuzziness by using the membership function; therefore, the sensitivity of fuzzy evaluation is quite high compared to other index evaluation techniques [20–22].

The following procedure describes fuzzy comprehensive assessment [21].

(a)   Selection of Factor Set U. Consider

$$U = \{u_i\}, \quad i = 1, 2, \ldots, n,$$

(1)

where n is the number of selected evaluation factors. In this study, 19 indicators listed in Table 1 are selected to build the factor set; in other words, $n = 19$.

(b)   Construction of Evaluation Criteria Set V. Consider

$$V = \{v_j\}, \quad j = 1, 2, \ldots, m,$$

(2)

where m is the number of evaluation criteria categories and V_j is the threshold of the jth criteria category. In the present study, outputs of the assessment are classified into five grades shown in Table 2, along with corresponding evaluation criteria of each indicator. From grade 1 to grade 5, the extent of groundwater environment influenced by tunnel construction ranges from very weak to very strong.

Table 2: Criteria for assessment of the negative effects caused by tunnel excavation on groundwater environment [16].

Indicator	Criteria				
	Very weak	Weak	Moderate	Strong	Very strong
[C_{11}] average annual rainfall (mm)	<600	600~800	800~1000	1000~1600	>1600
[C_{12}] average annual evaporation (mm)	>800	600~800	500~600	400~500	<400
[C_{13}] area of catchment zone (km²)	<5	5~10	10~30	30~50	>50
[C_{14}] coefficient of rainfall infiltration	<0.05	0.05~0.15	0.15~0.30	0.30~0.50	>0.50
[C_{15}] spatial relationship between the tunnel and geomorphology	Other (such as flat, protruding)	Flat and basin-shaped	Angular space and river crossing	Side below the valley and river crossing	Right below the valley and river crossing
[C_{16}] capacities of reservoirs and lakes on the ground (m³)	<1	1~10	10~50	50~300	>300
[C_{17}] flow of surface rivers (m³/s)	<0.1	0.1~0.5	0.5~2.0	2.0~10.0	>10.0
[C_{21}] carbonate rocks exposure ratio (%)	<30	30~50	50~70	70~90	>90
[C_{22}] water yield property of aquifers	<5	5~10	10~15	15~20	>20
[C_{23}] water pressure on the tunnel (Mpa)	<0.5	0.5~1.0	1.0~3.0	3.0~5.0	>5.0
[C_{24}] development of folds	No folds	Folds with undeveloped fissure	Folds with moderately developed fissure	Folds with developed fissure	Folds with developed faults
[C_{25}] development of fracture zones	Rarely developed	Poorly developed	Moderately developed	Developed	Well developed
[C_{26}] formation lithology	Mudstone, shale, or clay	Sandstone or fine sandstone	Granite or igneous rock	Metamorphic rock	Soluble rocks including limestone, dolomite, and so forth
[C_{27}] location of tunnel in horizontal and vertical hydrodynamic zoning of groundwater	Recharge area in horizontal and unsaturated zone in vertical zoning	Recharge area in horizontal and seasonal fluctuation zone in vertical zoning	Runoff area in horizontal and shallow saturation zone or deep circulation zone in vertical zoning	Runoff area in horizontal and stressful saturation zone in vertical zoning	Discharge area in horizontal zoning
[C_{31}] length of tunnel (km)	<1.0	1.0~3.0	3.0~10.0	10.0~30.0	>30.0
[C_{32}] area of disturbed range (m²)	<50	50~120	120~250	250~350	>350
[C_{33}] construction method	Tunnel boring machine method	New Austrian Tunneling Method	Partial excavation using drilling and blasting method	Benching tunneling using drilling and blasting method	Full face excavation using drilling and blasting method
[C_{34}] burial depth of tunnel (m)	Extremely bad for water inflow	Bad for water inflow	Moderate for water inflow	Good for water inflow	Extremely good for water inflow
[C_{35}] measures for prevention of groundwater flowing into tunnel	Composite lining and pregrouting	Composite lining and exterior waterproof (or postgrouting)	Composite lining	Structural self-waterproof	Drainage

(c)   Establishment of Membership Functions. In fuzzy logic, the set A is defined in terms of its membership function by

$$A = \{(f_A(x)), \ x \in X, \ f_A(x) \in [0,1]\},$$

(3)

where X is a domain, with a generic element of X denoted by x, f_A is the membership function of the set A, which maps the domain X onto the interval [0 1], and $f_A(x)$ represents the degree that x belongs to set A. x is a full member of A when $f(x) = 1$, not member of A when $f_A(x) = 0$, and a partial member of A when $f_A(x) = (0, 1)$. The membership function of each factor to the assessment criteria at each grade can be described quantitatively by a set of formulae as follows (4) (Note: if a big value represents a small contribution to water inrush and problems induced by it, then the direction of the inequality in the conditions should be reversed):

$$f_{ij}(x_i) = \begin{cases} 0 & x_i > v_{i(j+1)}, \\ \dfrac{\left(v_{i(j+1)} - x_i\right)}{\left(v_{i(k+1)} - v_{ik}\right)} & v_{ij} \leq x_i \leq v_{i(j+1)}, \quad j = 1, \\ 1 & x_i < v_{ij}, \end{cases}$$

$$f_{ij}(x_i) = \begin{cases} 0 & x_i > v_{i(j+1)}, \ x_i < v_{i(j-1)}, \\ \dfrac{\left(x_i - v_{i(j-1)}\right)}{\left(v_{ij} - v_{i(j-1)}\right)} & v_{i(j-1)} \leq x_i \leq v_{ij}, \\ \dfrac{\left(v_{i(j+1)} - x_i\right)}{\left(v_{i(j+1)} - v_{ij}\right)} & v_{ij} \leq x_i \leq v_{i(j+1)}, \end{cases}$$

$$(4)$$

where i is the number of evaluation factors ($i = 1, 2, 3$), j is the number of assessment criteria levels ($j = 1, 2, 3, 4, 5$), x_i is the actual value of evaluation factor i, V_{ij}, $V_i(j-1)$, $V_i(j+1)$ is the assessment criteria threshold of the ith assessment factor at level j, $j-1$, $j+1$, respectively, and $f_{ij}(x_i)$ is the membership degree of assessment factor i at level j.

(d) Calculation of Fuzzy Relation Matrix R. Substituting the data of each indicator and the gradation criteria into the membership function listed above, the fuzzy matrix R can be expressed as

$$R = \begin{bmatrix} r_{11} & r_{12} & \cdots & r_{1m} \\ r_{21} & r_{22} & \cdots & r_{2m} \\ \vdots & \vdots & \vdots & \vdots \\ r_{n1} & r_{n2} & \cdots & r_{nm} \end{bmatrix},$$

$$(5)$$

where r_{ij} ($i = 1, 2, \ldots, n; j = 1, 2, \ldots, m$) is the membership degree of the ith assessment parameter at the jth level.

(e)   Determination of Weight Set W. Consider

$$W = \{w_i\}, \quad i = 1, 2, \ldots, n, \tag{6}$$

where n is the number of selected evaluation factors and w_i is the weight of ith factor indicating the relative importance. Determination of the relative weight of each indicator is important for making an appropriate assessment. 10 experts and professors engaged in hydrological and environmental studies coming from home and broad were invited to identify the relative importance of each category and indicator according to their knowledge and experience. After obtaining the sequence and its score representing the relative importance of each factor, weight set (Table 3) is calculated using G1-rank correlation analysis method proposed by Guo [23].

Table 3: Comprehensive weights of the indicators included in the indicator system [16].

Indicator	C_{11}	C_{12}	C_{13}	C_{14}	C_{15}	C_{16}	C_{17}	C_{21}	C_{22}	C_{23}
Weight	0.0535	0.0251	0.0387	0.0386	0.0539	0.0419	0.0419	0.0679	0.0705	0.0393
Indicator	C_{24}	C_{25}	C_{26}	C_{27}	C_{31}	C_{32}	C_{33}	C_{34}	C_{35}	
Weight	0.0617	0.0834	0.0781	0.0681	0.0554	0.0362	0.0477	0.0445	0.0538	

(f)   Fuzzy Matrix Composition and Determination of the Final Evaluation Result. Fuzzy composition evaluation can be performed as follows:

$$E = W * R = (w_1, w_2, \ldots, w_n) \begin{bmatrix} r_{11} & \cdots & r_{15} \\ \vdots & \ddots & \vdots \\ r_{n1} & \cdots & r_{n5} \end{bmatrix}$$

$$= (e_1, e_2, \ldots, e_5), \tag{7}$$

where W and R are the weight set and the fuzzy relationship matrix determined above, respectively, and $*$ is a fuzzy composite operator which is very critical to final evaluation results. In this study, average fuzzy composite operator is chosen.

CASE STUDY OF TONGLUOSHAN TUNNEL

General Description

Project Profile

Tongluoshan tunnel, with a length about 5.2 km, located in Guang'an, Southwest China (Figure 1), is a key project of Dianjiang-Linshui expressway. This project consists of two parallel tunnels, between which the distance is 30 m (from the adjacent walls). The thickest overburden of Tongluoshan tunnel is approximately 280 m, while the minimum value is less than 40 m. In order to save time, excavations were executed simultaneously from the northwest portal and southeast portal to the center during 2005 to 2008.

Figure 1: Location map of Tongluoshan tunnel.

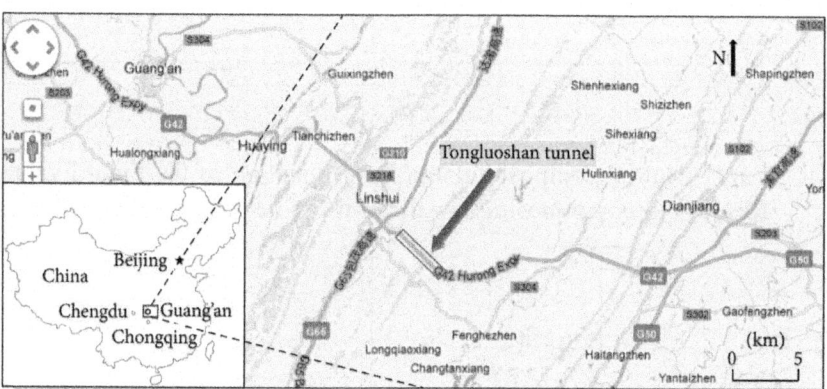

Climatic and Hydrological Features of the Tunnel Area

The tunnel area belongs to subtropics monsoon climate region, where precipitation is abundant but distributing unevenly throughout the year with the majority (70%) falling from May to September. Annual precipitation in the tunnel area oscillates between 836.6 mm and 1529.8 mm with an average of 1215.5 mm. Mean annual temperature in the tunnel area is 16.9°C, with the highest and lowest daily temperature 40.4°C and −3.8°C, respectively.

Tongluoshan tunnel is located in the middle of Tongluoshan anticline, where Yulin River and Zhonghe River, belonging to the secondary and first branch of Yangtze River respectively, are the main regional surface water bodies. The major river in the tunnel area is Qingshuixi River, which originates from Jinzhong reservoir northeast of Tongluoshan tunnel and flows approximately

8.8 km along the axis of Tongluoshan anticline and then turns to southeast. After about 2.5 km, it becomes a part of Zhonghe River. As is recorded, the average annual flow rate of Qingshuixi River is about 0.355 m^3/s before tunnel construction.

Geological and Hydrogeological Characteristics of the Tunnel Area

Ground surface elevation ranges from 156 m to 1053 m in a regional scale while the northern part is higher than the southern part which Yangtze River goes through. In the tunnel area, elevation in the valley along the route of Tongluoshan anticline oscillates between 400 m to 550 m, which is much smaller than that in the anticlinal flanks ranging from 600 m to 750 m.

According to the preliminary geological survey undertaken by Sichuan Institute of Coal Field Geological Engineering Exploration and Design [24] and other information about tunnel design, few except a small fault (F_1) and Tongluoshan anticline are the major geological structures developing in the tunnel area. And the stratigraphic sequence consists of carbonate rocks of lower to middle Triassic (T_2, T_1), coal measure strata of upper Triassic (T_3), mudstone interbedded with siltstone of lower to middle Jurassic (J_2, J_1), and unconsolidated sediments in Quaternary (Q). The geological profile along the route is shown in Figure 2. Tongluoshan tunnel is located in the southern part of Qingshuixi secondary hydrogeological unit, which is bounded by Yujiayakou watershed in the south, Shengouzigou watershed in the north, Yulin River in the west, and Zhonghe River in the east. This hydrogeological unit has an area about 95 km^2, stretching 11.9 km from the southern boundary to the northern boundary, and 7 to 9 km in the west-east direction. According to the topography, geological structure, water-bearing medium and recharge, runoff, and discharge condition of groundwater in this hydrogeological unit, it can be divided into four aquifer systems, composed of pore water in the unconsolidated formation of Quaternary, pore and fissure water in the clasolite of Jurassic, pore and fissure water in the clasolite of upper Triassic, and carbonate water in the carbonate rocks of lower to middle Triassic. Precipitation is believed to be the primary source of recharge to the aquifers. Groundwater recharge occurs mainly from infiltration of precipitation into outcropping carbonate rocks, including stratum of lower to middle Triassic principally distributed in the core area of Tongluoshan anticline. Controlled by the coal-bearing strata and mudstone interbedded with siltstone located in the anticlinal flanks, karst groundwater generally flows towards the south and mainly discharges in terms of springs along Qingshuixi River, which is viewed as local base level of erosion.

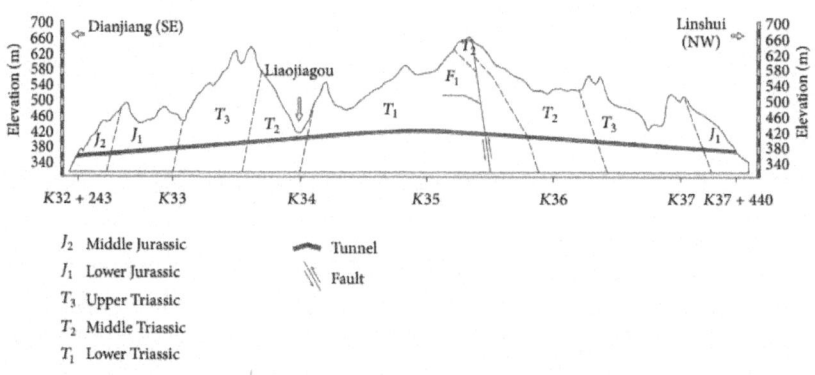

Figure 2: Geological profile of Tongluoshan tunnel.

Fuzzy Comprehensive Evaluation of the Influence Caused by Tunnel Excavation on Groundwater Environment

Quantification of the Indicators

In order to assess the negative effects caused by Tongluoshan tunnel construction on groundwater environment, the value of each indicator included in the indicator system should be previously quantified based on field investigation and technological reports related to the project such as geological survey undertaken by Sichuan Institute of Coal Field Geological Engineering Exploration and Design [24] and so on. As some indicators, for example, lithology of the formation and relation between tunnel and topography and burial depth of tunnel, are unique and of great importance to the evaluation result that may play a vital role in decision making, the work of quantification of the indicators should be done as carefully as it can be, aiming at assessing the influence as accurate as possible. Specific value representing quantification of each indicator is listed in Table 4.

Table 4: Quantification of each factor included in the indicator system.

Category	Indicator	Value of Tonluoshan tunnel
	[C_{11}] average annual rainfall (mm)	According to rainfall records from local meteorological station, the average annual precipitation is 1215.5 mm in the tunnel area.
	[C_{12}] average annual evaporation (mm)	According to evaporation records from local meteorological station, the average annual evaporation is 959.6 mm in the tunnel area.
[B_1] physical geography	[C_{13}] area of catchment zone (km²)	Tongluoshan tunnel passes through strata consisting of carbonate rocks of high permeability in the core and clastic rocks of low permeability in the flanks of Tongluoshan anticline. When calculating the catchment zone of tunnel inflow, a region composed of local watershed around the route of tunnel in the flanks and the entire karst valley are included. Total area of the catchment zone of Tongluoshan tunnel then is determined to 38.4 km².
	[C_{14}] coefficient of rainfall infiltration	Based on the geological investigation report, infiltration coefficient of rainfall in the outcropping carbonate rocks (T_1 and T_2) reaches 0.55, with 0.20 in the coal measure strata (T_3), while it is only 0.054 in the Jurassic clastic rocks (J_1 and J_2). Due to statistics of outcrop in terms of different lithology, 44% of them are carbonate rocks, while 32% are coal bearing strata, and 24% are clastic rocks.
	[C_{15}] spatial relationship between the tunnel and geomorphology	Length consisting of the section between 32 K + 700 and 34 K + 200 and that between 35 K + 800 and 36 K + 300 is about 2 km, belonging to the type of side below the valley and river crossing, while the rest belongs to other type.
	[C_{16}] capacities of reservoirs and lakes on the ground (m³)	There are no reservoirs and lakes within 2 km from the tunnel axis except Jinzhong reservoir, which is about 9 km far from the tunnel axis, storing approximately 300 thousand cubic meters. Since karstification is well developed in the tunnel area and Jinzhong reservoir is the origin of Qingshuixi River, it is taken into consideration from a safe point of view.
	[C_{17}] flow of surface rivers (m³/s)	Qingshuixi River is the main surface river which may have hydraulic connection with the water inflow into tunnel.
	[C_{21}] carbonate rocks exposure ratio (%)	As mentioned above, this ratio is about 44% in the tunnel area.
	[C_{22}] water yield property of aquifers	Based on the longitudinal profile of Tongluoshan tunnel, area of rocks with poor water yield property occupies 62.37%, with middle standing 30.14%, while the left (7.5%) is considered as aquifer with very good water yield property.
[B_2] geology and hydrogeology	[C_{23}] water pressure on the tunnel (Mpa)	Average water pressure on the tunnel is estimated as 1.0 MPa.
	[C_{24}] development of folds	Tongluoshan anticline is the main fold developed in the tunnel area with fractures coming from geological and karstic process.
	[C_{25}] development of fracture zones	Fracture zone passed through by tunnel excavation is found moderately developed.
	[C_{26}] formation lithology	Based on outcrops in the tunnel area, mudstone, shale, and clay stand for 21% and sandstone and siltstone stand for 35%, while carbonate rocks occupy 44%.
	[C_{27}] location of tunnel in horizontal and vertical hydrodynamic zoning of groundwater	From a regional scale, Tongluoshan tunnel is located in runoff area in horizontal and stressful saturation zone in vertical zoning.
	[C_{31}] length of tunnel (km)	About 5.2 km.
	[C_{32}] area of disturbed range (m²)	About 185 m² including two tunnels.
[B_3] tunnel engineers	[C_{33}] construction method	About 80% of the tunnel excavated by full face excavation using drilling and blasting method, while the others adopt benching tunneling using drilling and blasting method.
	[C_{34}] burial depth of tunnel (m)	Burial depth between 100 m and 300 m stands for 50%, while the others have a value less than 100 m. From some statistic data and hydrogeological condition in the tunnel area, moderate level is given.
	[C_{35}] measures for prevention of groundwater flowing into tunnel	Composite lining is the prevailing waterproof used by Tongluoshan tunnel, while 20% of which adopts external pregrouting.

Data Preprocessing

Data preprocessing for comprehensive evaluation is to choose an appropriate membership function which determines the membership of each indicator according to given criteria. In this case study, trapezoidal function and semitrapezoidal function (4) are chosen, because of their advantage in simple processing and smooth linking. It is noted that the standard of an indicator at each level can be described as an interval like $[a, b]$, which may have existed or will be built. And it is reasonable for determining the membership of x as 1.0 when $x = (a + b)/2$, so V_{ij} in the equations (e.g., (4)) may be valued as $(a + b)/2$. If difficulties existing in finding an upper or lower limit of some indicators, $\eta = 3\sim5$ or $\eta = 1/5\sim1/3$ can be multiplied to achieve this goal. After substitution of actual value quantified in Table 4 and criteria listed in Table 2 into the membership function, membership of each indictor at every level (Table 5) can be gotten, which will directly be transferred into a fuzzy matrix R.

Table 5: Membership of each indictor at every level calculated from quantification of Tongluoshan tunnel and the evaluation criteria.

Indicator	Grades				
	Very weak	Weak	Moderate	Strong	Very strong
$[C_{11}]$ average annual rainfall (mm)	0	0	0.211	0.789	0
$[C_{12}]$ average annual evaporation (mm)	0.288	0.712	0	0	0
$[C_{13}]$ area of catchment zone (km²)	0	0	0.080	0.920	0
$[C_{14}]$ coefficient of rainfall infiltration	0.158	0.146	0.256	0.330	0.110
$[C_{15}]$ spatial relationship between the tunnel and geomorphology	0.600	0	0	0.400	0
$[C_{16}]$ capacities of reservoirs and lakes on the ground (m³)	0	0	1.000	0	0
$[C_{17}]$ flow of surface rivers (m³/s)	0	1.000	0	0	0
$[C_{21}]$ carbonate rocks exposure ratio (%)	0	0.800	0.200	0	0
$[C_{22}]$ water yield property of aquifers	0.624	0	0.301	0	0.075
$[C_{23}]$ water pressure on the tunnel (Mpa)	0	0.800	0.200	0	0
$[C_{24}]$ development of folds	0	0	0	1.000	0
$[C_{25}]$ development of fracture zones	0	0	1.000	0	0
$[C_{26}]$ formation lithology	0.210	0.350	0	0	0.440
$[C_{27}]$ location of tunnel in horizontal and vertical hydrodynamic zoning of groundwater	0	0	0	1.000	0
$[C_{31}]$ length of tunnel (km)	0	0.289	0.711	0	0
$[C_{32}]$ area of disturbed range (m²)	0	0	1.000	0	0
$[C_{33}]$ construction method	0	0	0	0.200	0.800
$[C_{34}]$ burial depth of tunnel (m)	0	0	1.000	0	0
$[C_{35}]$ measures for prevention of groundwater flowing into tunnel	0.200	0	0.800	0	0

Fuzzy Comprehensive Evaluation

When prepared, evaluation set W and fuzzy matrix R can be composed using average fuzzy composite operator according to (7):

$$E = W * R = (w_1, w_2, \ldots, w_{19}) \begin{bmatrix} r_{1,1} & \cdots & r_{1,5} \\ \vdots & \ddots & \vdots \\ r_{19,1} & \cdots & r_{19,5} \end{bmatrix}$$

$$(8)$$

= (0.117, 0.195, 0.355, 0.251, 0.082).

Maximum membership principle [25, 26] is a simple and widely used principle on the membership degree matrix. The elements in the vector of evaluation result stand for the membership degree to assessing level. According to this principle, it can be determined that the assessment result of Tongluoshan tunnel is moderate, as 0.355 is the maximum member in the evaluation result vector. Meanwhile, another weighted mean method [27] which allocates a relative rating (1, 2, 3, 4, and 5, resp.) to the five levels and uses accelerations composition method to calculate the total value is employed to perform calculations:

$$T = 1 \times e_1 + 2 \times e_2 + 3 \times e_3 + 4 \times e_4 + 5 \times e_5$$

$$= 1 \times 0.117 + 2 \times 0.195 + 3 \times 0.355 + 4 \times 0.251 + 5$$

$$\times 0.082$$

$$= 2.99. \tag{9}$$

Based on the result calculated by weighted mean method, middle level tends to be accepted because T is very close to 3. Therefore, from arithmetically speaking, moderate grade is recommended as the influence level of Tongluoshan tunnel, indicating that the local groundwater environment might have suffered a medium degree of impact from the tunnel excavation. Although the indicator system supplying us with a convenient way to evaluate the influence resulted from tunnel construction on groundwater environment, field work such as environmental investigation and hydrological monitoring is important and essential to help judge whether the assessment is appropriate or not. Some details about the procedure and result of environmental investigation and hydrological monitoring carried out in Tongluoshan tunnel area are described in the next section.

Environmental Investigation and Hydrologic Monitoring

Environmental investigation and hydrologic monitoring in the project area that may be affected was implemented for the early detection of environmental impacts caused by tunnel excavation. More than fifty springs, wells, and streams were found in the tunnel area during the first investigation carried in July 2005, and fifteen of them including twelve springs, one well, and two streams, which either supply people with drinking water or flow at a notable rate, were finally selected to perform a long term monitoring program. In addition, discharge from tunnel and rainfall in the project area were monitored together. Figure 3 shows the location of the observation points included in the frequent sampling protocol.

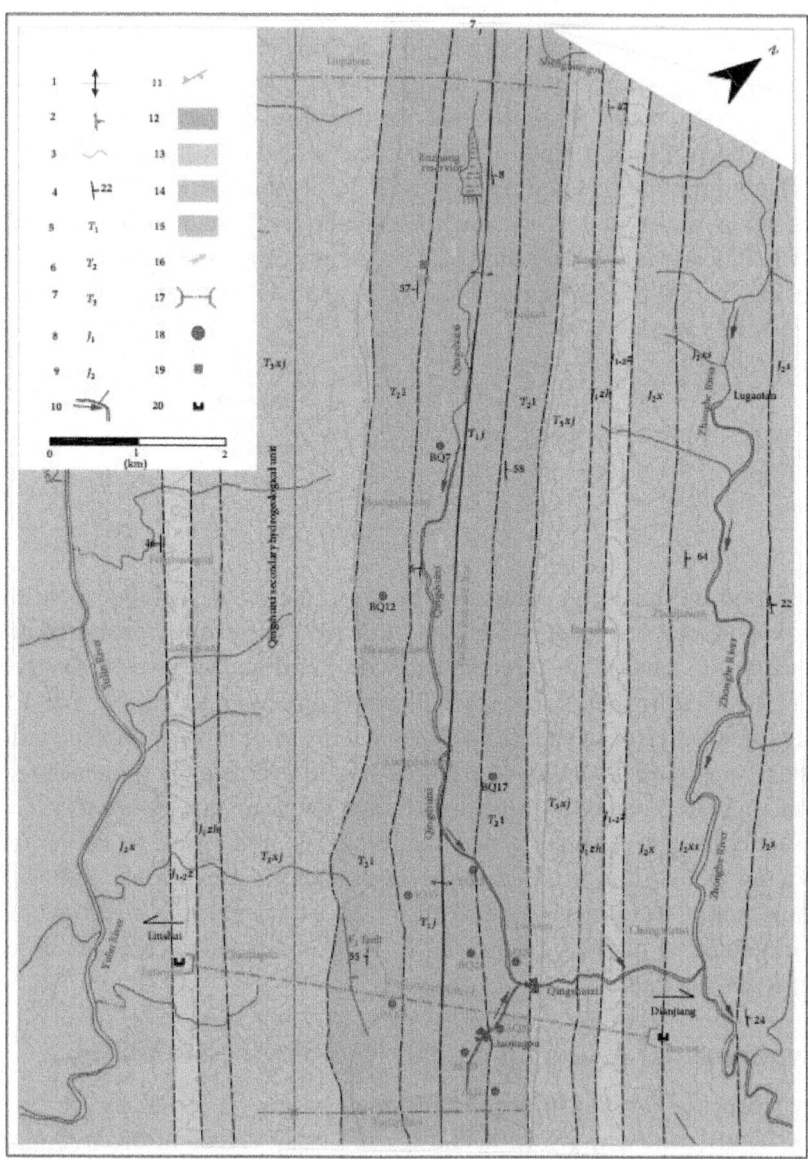

Figure 3: Map showing hydrogeological features and observation points in the project area. 1: anticline; 2: fault; 3: stratigraphic boundary; 4: attitude of strata; 5: lower Triassic; 6: middle Triassic; 7: upper Triassic; 8: lower Jurassic; 9: upper Jurassic; 10: river and its flow direction; 11: watershed; 12: limestone and dolomite; 13: carbonate rocks interbedded with mudstone; 14: sandstone; 15: mudstone interbedded with siltstone; 16: flow direction of groundwater; 17: tunnel; 18: spring; 19: well; 20: drainage from tunnel.

The monitoring process, lasting about 15 months from May 2006 to July 2007, was executed within the main construction period, during which daily discharge rate of tunnel and springs, flow of streams, and water table in well BJ1 were observed. In general, almost all the springs and streams as well as BJ1 respond to rainfall, with more responses from BJ1, BQ7, BQ27, AQ23, AQ37, Qingshuixi, and Liaojiagou, followed by BQ12, BQ25, AQ9, and AQ26. Another type including AQ21 and AQ35 shows evident time lag between discharge of springs and the precipitation, indicating that the fracture network recharging from rainfall and transferring groundwater to the emergence place does not work efficiently. More surprising, there are some springs which become nearly of no interest in precipitation, such as BQ17 and BQ26.

From a water balance point of view, it is believed that groundwater discharge from tunnel acting as a new flow out pattern will reduce the amount of other flow-out patterns such as springs and wells, supposing no changes happening to the flow in patterns. Consequently, hydrograph of surface water and groundwater can help us to identify the influence caused by tunnel construction.

AQ23, BQ25, and BQ27 are karst springs which have never dried up in the past according to local residents, appearing in stratum. It is obviously described in Figure 4(a) that these springs respond quickly to rainfall, especially in some periods with sufficient recharge. However, unfortunate things happen to AQ23 and BQ25 during the monitoring period. AQ23 completely dried up during June 27 to 30, 2006, July 29 to September 4, 2006, September 16 to 27, 2006, December 22, 2006, to January 11, 2007, and March 15 to April 1, 2007. BQ25 had no discharge during July 23 to September 27, 2006, October 15 to November 28, 2006, December 10, 2006, to January 10, 2007, January 18 to February 12, 2007, February 23 to April 1, 2007, and May 1 to 31, 2007. These phenomena indicate that Tongluoshan tunnel may drain groundwater which could have come to surface from AQ23 and BQ25. Compared to them, BQ27 has more luck because it had never dried up despite the fact that the minimum flow rate was about 0.1 L/s, far below the normal average 0.8 L/s.

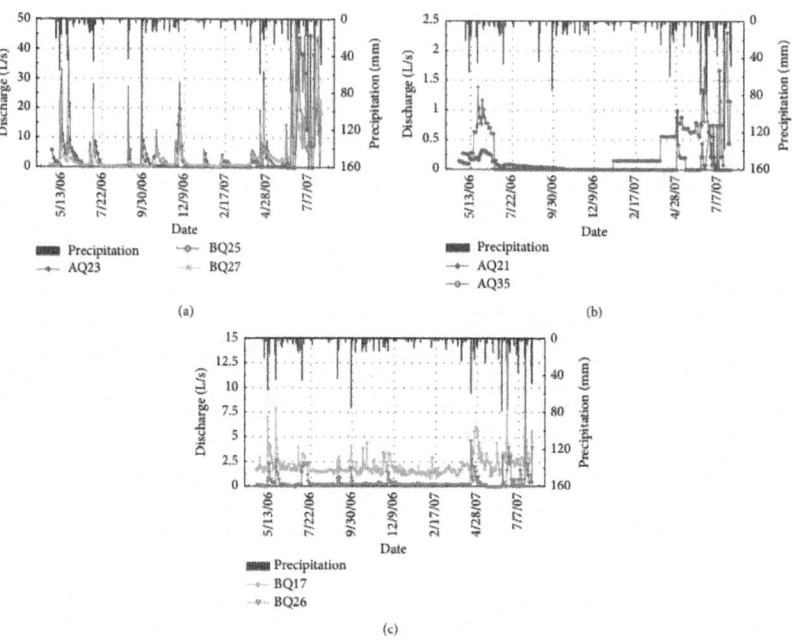

Figure 4: Hydrographs of the principal springs in the project area. (a) Discharge of AQ23, BQ25, and BQ27 versus time. (b) Discharge of AQ21 and AQ35 versus time. (c) Discharge of BQ26 and BQ27 versus time.

AQ21 flowed out from an abandoned mine, which settled in coal bearing layer. It continuously supplied people around with groundwater for drinking prior to tunnel excavation. According to the records shown in Figure 4(b), the flow came to zero during June 22, 2006, to January 9 and July 1 to 31, 2007, in spite of adequate rainfall during January 10 to June 30, 2007. AQ35 is an epikarst spring situated in higher elevation, as is recorded in Figure 4(b); its discharge declined from 0.3 L/s to zero using approximately 4 months and maintains this status for a long time regardless of whether rainfall occurred. Based on the situation described above, it is doubted that AQ21 and AQ35 may had been linked to tunnel construction. According to Figure4(c), BQ17 flowed throughout the period recorded with modest discharge, even during rainy periods. This phenomenon can be attributed to the abundant storage capacity of fractures that feeding BQ17.

Liaojiagou is a branch of Qingshuixi, which is the major stream in the project area. According to Figures 5 and4(a), both Qingshuixi and Liaojiagou bear much resemblance to AQ23, BQ25, and BQ27 in response to rainfall. But Liaojiagou, as with AQ23, did not escape influences from both local meteorological condition and tunnel excavation. They dried up quickly after being recharged from precipitation and continued this state for a long period. However, Qingshuixi showed strong vitality during the whole monitoring period, flowing in quite accordance with precipitation recorded in the project area without any appearance of external leakage, demonstrating that Qingshuixi has not been affected by construction of Tongluoshan tunnel.

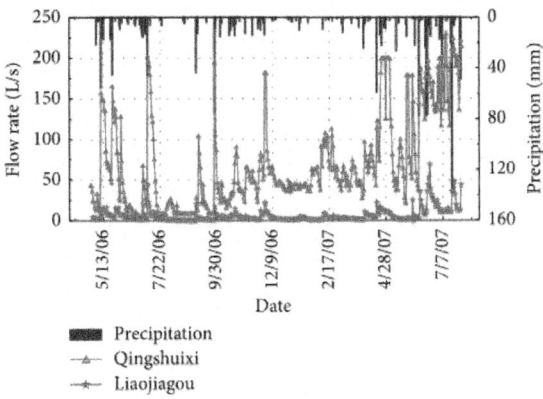

Figure 5: Hydrographs of Qingshuixi and Liaojiagou.

Drainage from Tongluoshan tunnel is depicted in Figure 6, from which we can find a quick increase from 20 L/s to 64 L/s before June 28, 2006, and drastic oscillation between 32 L/s and 69 L/s in the next three months, followed by pediocratic changes between 30 L/s and 40 L/s. In general, tunnel inflow had an obvious ascending tendency when encountering carbonate rocks and performed a slow downtrend after drainage of groundwater stored in the fracture network around tunnel and grouting to prevent water inrush.

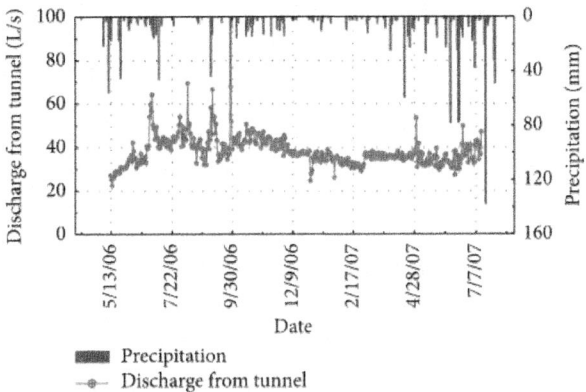

Figure 6: Drainage from tunnel versus time.

Determination of an Impact

While employing the indicator system to assess the negative effects caused by Tongluoshan tunnel construction on groundwater environment, maximum membership principle method makes an assessment result of medium level as 0.355 is the maximum member in the evaluation result vector, and weighted mean method gives nearly the same result because T is very close to 3. From a mathematic point of view, moderate grade is recommended as the influence level of Tongluoshan tunnel.

In order to verify whether the assessment is reasonable or not, environmental investigation and hydrological monitoring program was carried out in the tunnel area. Twelve springs, one well, and two streams, which either supply people with drinking water or flow at a notable rate, were selected to perform a 15-month-long monitoring process and so were the discharge from tunnel and rainfall in the project area. During the main construction period, some of the monitoring points near the tunnel axis, such as AQ21, AQ23, AQ35, and BQ27, have been strongly impacted by tunnel construction and lasting drainage of groundwater. Most of them have never dried up and maintained base flow even in the dry season, but an obvious decline was found throughout the monitoring period. Though strong relation exists between most of the monitoring points and precipitation, it seems that groundwater drainage from tunnel plays a nonnegligible role because some of the monitoring points repeatedly ran dry during the main construction period despite being recharged by precipitation. Based on the information fed back from environmental investigation and hydrologic monitoring carried out during the main construction period, approximately 1 km from the tunnel axis (not including BQ25 which is partly fed by upstream surface water) is inferred as the sphere of influence in the middle karst section of Tongluoshan tunnel, which is less than other typical karst tunnel, such as Yuanliangshan tunnel and Geleshan tunnel [8, 16].

As is analyzed above, moderate grade influence on groundwater system is considered to be an appropriate result for Tongluoshan tunnel construction for the reason that (a) assessment made from the indicator system and fuzzy comprehensive method falls into middle level; (b) some of the monitoring points near the tunnel axis which have never dried up and maintained base flow even in the dry season experienced an obvious decline throughout the monitoring period, revealing that the local groundwater environment might have been impacted by the tunnel excavation; (c) the sphere of influence in the middle karst section of Tongluoshan tunnel approaches approximately 1 km from the tunnel axis, which is less than other typical karst tunnels. So, there is no doubt that the local groundwater environment had been impacted by the excavation of Tongluoshan tunnel, but the influence level is moderate.

CONCLUSION AND OUTLOOK

Making an assessment on how the construction would affect local groundwater environment before excavation is of vital importance to line selection of tunnel. This study adopted an indicator system and fuzzy comprehensive evaluation method to identify the influence caused by Tongluoshan tunnel construction on groundwater environment. It was shown that maximum membership stood in the middle of the evaluation result set and 3.2 was scored by relative rating weighted calculation, indicating that excavation of Tongluoshan tunnel had influenced the local groundwater environment at a moderate degree. Based on the information fed back from environmental investigation and hydrologic monitoring carried out during the main construction period, flow of some springs and Liaojiagou stream located beside the tunnel route was found experiencing an apparent decline; even worse, AQ21, AQ23, AQ35, and BQ25 had dried up for a relatively long time because of lower precipitation during 2006 than that in the past and drainage from tunnel resulting in reduction of other flow-out patterns in water balance.

Through this practice, it can be concluded that making an assessment using the indicator system and fuzzy comprehensive evaluation method would supply us with a simple way to understand the situation that may happen to groundwater environment during tunnel construction and can help us to choose an optimized channel through which less disruption related to water inflow may occur. Once the tunnel location is determined and prepared to construction, we can also use the indicator system to evaluate the potential groundwater environmental impact before building it and take appropriate measures to minimize the impact. The major differences between a building tunnel and a built tunnel focus on two indicators including "construction method" and "measures for prevention of groundwater flowing into tunnel." Taking Tongluoshan tunnel as an example and setting "boring machine method" and "drainage with composite lining" as the initial condition, a result vector of 0.106, 0.195, 0.312, 0.242, and 0.145 indicates that the influence level partly tends to be strong, because the gap between the maximum member and the second maximum member falls to 0.07, smaller than 0.1. In order to minish the potential impact, measurements such as previous and postgrouting from tunnel, pregrouting from surface, and partial excavation using drilling and blasting method were taken at some sections with high risk of water irruption, reducing the influence level and impact on groundwater environment to some extent.

ACKNOWLEDGMENTS

This research was supported by Communications Department of Sichuan Province, Sichuan Dayu Expressway Construction and Development Company Limited (2006-G-02), and the Fundamental Research Funds for the Central Universities (2682015CX020). Many thanks should be given to the team members related to this program, for their help with field investigation and data collection and processing.

REFERENCES

1. H. R. Zarei, A. Uromeihy, and M. Sharifzadeh, "Identifying geological hazards related to tunneling in carbonate karstic rocks-Zagros, Iran," Arabian Journal of Geosciences, vol. 5, no. 3, pp. 457–464, 2012.

2. C. Butscher, P. Huggenberger, and E. Zechner, "Impact of tunneling on regional groundwater flow and implications for swelling of clay-sulfate rocks," Engineering Geology, vol. 117, no. 3-4, pp. 198–206, 2011.

3. U. Chiocchini and F. Castaldi, "The impact of groundwater on the excavation of tunnels in two different hydrogeological settings in central Italy," Hydrogeology Journal, vol. 19, no. 3, pp. 651–669, 2011.

4. J.-H. Hwang and C.-C. Lu, "A semi-analytical method for analyzing the tunnel water inflow," Tunnelling and Underground Space Technology, vol. 22, no. 1, pp. 39–46, 2007.

5. S. Alija, F. J. Torrijo, and M. Quinta-Ferreira, "Geological engineering problems associated with tunnel construction in karst rock masses: the case of Gavarres tunnel (Spain)," Engineering Geology, vol. 157, pp. 103–111, 2013.

6. M. J. Day, "Karstic problems in the construction of Milwaukee's Deep Tunnels," Environmental Geology, vol. 45, no. 6, pp. 859–863, 2004.

7. L. Schwarz, I. Reichl, H. Kirschner, and K. P. Robl, "Risks and hazards caused by groundwater during tunnelling: geotechnical solutions used as demonstrated by recent examples from Tyrol, Austria,"Environmental Geology, vol. 49, no. 6, pp. 858–864, 2006. · ·

8. L. W. Jiang, Y. J. Yi, and Z. M. Jia, "Research on characteristics and formation mechanism of great deep buried infilled caves at Maoba syncline in the Yuanliangshan railway tunnel," Journal of Railway Engineering Society, no. 4, pp. 53–60, 2007 (Chinese).

9. D.-J. Tseng, B.-R. Tsai, and L.-C. Chang, "A case study on ground treatment for a rock tunnel with high groundwater ingression in Taiwan," Tunnelling and Underground Space Technology, vol. 16, no. 3,

pp. 175–183, 2001.

10. C. Zangerl, E. Eberhardt, and S. Loew, "Ground settlements above tunnels in fractured crystalline rock: numerical analysis of coupled hydromechanical mechanisms," Hydrogeology Journal, vol. 11, no. 1, pp. 162–173, 2003.

11. J. Gisbert, A. Vallejos, A. González, and A. Pulido-Bosch, "Environmental and hydrogeological problems in karstic terrains crossed by tunnels: a case study," Environmental Geology, vol. 58, no. 2, pp. 347–357, 2009.

12. V. Vincenzi, A. Gargini, and N. Goldscheider, "Using tracer tests and hydrological observations to evaluate effects of tunnel drainage on groundwater and surface waters in the Northern Apennines (Italy)," Hydrogeology Journal, vol. 17, no. 1, pp. 135–150, 2009.

13. P. M. Attanayake and M. K. Waterman, "Identifying environmental impacts of underground construction," Hydrogeology Journal, vol. 14, no. 7, pp. 1160–1170, 2006.

14. J. R. Raposo, J. Molinero, and J. Dafonte, "Quantitative evaluation of hydrogeological impact produced by tunnel construction using water balance models," Engineering Geology, vol. 116, no. 3-4, pp. 323–332, 2010.

15. F.-R. Yang, C.-H. Lee, W.-J. Kung, and H.-F. Yeh, "The impact of tunneling construction on the hydrogeological environment of 'Tseng-Wen Reservoir Transbasin Diversion Project' in Taiwan,"Engineering Geology, vol. 103, no. 1-2, pp. 39–58, 2009.

16. J. Liu, Study on the evaluation system of negative effects on groundwater environment resulted by tunnel construction in karst area [Ph.D. thesis], Southwest Jiaotong University, Chengdu, China, 2011, (Chinese).

17. D. X. Yuan, Karstology of China, Geological Publishing House, Beijing, China, 1994, (Chinese).

18. J. F. Durand, "The impact of gold mining on the Witwatersrand on the rivers and karst system of Gauteng and North West Province, South Africa," Journal of African Earth Sciences, vol. 68, no. 15, pp. 24–43, 2012.

19. L. A. Zadeh, "Fuzzy sets," Information and Control, vol. 8, pp. 338–353, 1965.

20. G. Onkal-Engin, I. Demir, and H. Hiz, "Assessment of urban air quality in Istanbul using fuzzy synthetic evaluation," Atmospheric Environment, vol. 38, no. 23, pp. 3809–3815, 2004

21. C. Q. Mi, X. D. Zhang, S. M. Li, J. Yang, D. Zhu, and Y. Yang,

"Assessment of environment lodging stress for maize using fuzzy synthetic evaluation," Mathematical and Computer Modelling, vol. 54, no. 3-4, pp. 1053–1060, 2011.

22. S. Dahiya, B. Singh, S. Gaur, V. K. Garg, and H. S. Kushwaha, "Analysis of groundwater quality using fuzzy synthetic evaluation," Journal of Hazardous Materials, vol. 147, no. 3, pp. 938–946, 2007.

23. Y. J. Guo, "New theory and method of dynamic comprehensive evaluation," Journal of Management Sciences, vol. 5, no. 2, pp. 52–57, 2002 (Chinese).

24. Sichuan Institute of Coalfield Geological Engineering Exploration and Designing, Preliminary Geological Survey of Tongluoshan Tunnel, Sichuan Institute of Coalfield Geological Engineering Exploration and Designing, Chengdu, China, 2003.

25. Y. Wang, W. F. Yang, M. Li, and X. Liu, "Risk assessment of floor water inrush in coal mines based on secondary fuzzy comprehensive evaluation," International Journal of Rock Mechanics & Mining Sciences, vol. 52, pp. 50–55, 2012.

26. Q. Zhang, X. Yang, Y. Zhang, and M. Zhong, "Risk assessment of groundwater contamination: a multilevel fuzzy comprehensive evaluation approach based on DRASTIC model," The Scientific World Journal, vol. 2013, Article ID 610390, 9 pages, 2013.

27. H. Y. Wu, K. L. Chen, Z. H. Chen et al., "Evaluation for the ecological quality status of coastal waters in East China Sea using fuzzy integrated assessment method," Marine Pollution Bulletin, vol. 64, no. 3, pp. 546–555, 2012.

CITATION

CHAPTER 1

Shirdel, B. (2015) Analysis of Engineering Geology Indices in Three Units of Atamir Formation in Kope Dagh Zone. Open Journal of Geology, 5, 847-875. doi: 10.4236/ojg.2015.512072.

CHAPTER 2

Selim, S. , Hamdan, A. and Rady, A. (2014) Groundwater Rising as Environmental Problem, Causes and Solutions: Case Study from Aswan City, Upper Egypt. *Open Journal of Geology*, **4**, 324-341. doi: 10.4236/ojg.2014.47025.

CHAPTER 3

Merayyan, S. and Safi, S. (2014) Feasibility of Groundwater Banking under Various Hydrologic Conditions in California, USA. Computational Water, Energy, and Environmental Engineering, 3, 79-92. doi: 10.4236/cweee.2014.33009.

CHAPTER 4

S. Mangala PRAVEENA, M. Harun ABDULLAH, A. Zaharin ARIS and K. BIDIN, "Groundwater Solution Techniques: Environmental Applications," *Journal of Water Resource and Protection*, Vol. 2 No. 1, 2010, pp. 8-13. doi:10.4236/jwarp.2010.21002.

CHAPTER 5

Djémin, J. , Kouamé, J. , Deh, K. , Abinan, A. and Jourda, J. (2016) Contribution of the Sensitivity Analysis in Groundwater Vulnerability Assessing Using the DRASTIC Method: Application to Groundwater in Dabou Region (Southern of Côte d'Ivoire). *Journal of Environmental Protection*, **7**, 129-143. doi: 10.4236/jep.2016.71012.

CHAPTER 6

Rebelo, M. , Santos, G. and Silva, R. (2014) A Methodology to Develop the Integration of the Environmental Management System with Other Standardized Management Systems. Computational Water, Energy, and Environmental Engineering, 3, 170-181. doi: 10.4236/cweee.2014.34018.

CHAPTER 7

Yazdi, A. , Arian, M. and Tabari, M. (2014) Geological and Geotourism Study of Iran Geology Natural Museum, Hormoz Island. *Open Journal of Ecology*, **4**, 703-714. doi: 10.4236/oje.2014.411060.

CHAPTER 8

L. Khodapanah, W. Sulaiman and H. Nassery, "Hydrogeological Framework and Groundwater Balance of a Semi-arid Aquifer, a Case Study from Iran," *Journal of Water Resource and Protection*, Vol. 3 No. 7, 2011, pp. 513-521. doi: 10.4236/jwarp.2011.37061.

CHAPTER 9

Arshad Ashraf (2013). Changing Hydrology of the Himalayan Watershed, Current Perspectives in Contaminant Hydrology and Water Resources Sustainability, Dr. Paul Bradley (Ed.), ISBN: 978-953-51-1046-0, InTech, DOI: 10.5772/54492.

CHAPTER 10

Luc Descroix, Ibrahim Bouzou Moussa, Pierre Genthon, Daniel Sighomnou, Gil Mahé, Ibrahim Mamadou, Jean-Pierre Vandervaere, Emmanuèle Gautier, Oumarou Faran Maiga, Jean-Louis Rajot, Moussa Malam Abdou, Nadine Dessay, Aghali Ingatan, Ibrahim Noma, Kadidiatou Souley Yéro, Harouna Karambiri, Rasmus Fensholt, Jean Albergel and Jean-Claude Olivry (2013). Impact of Drought and Land – Use Changes on Surface – Water Quality and Quantity: The Sahelian Paradox, Current Perspectives in Contaminant

Hydrology and Water Resources Sustainability, Dr. Paul Bradley (Ed.), ISBN: 978-953-51-1046-0, InTech, DOI: 10.5772/54536.

CHAPTER 11

Wei-Zu Gu, Jiu-Fu Liu, Jia-Ju Lu and Jay Frentress (2013). Current Challenges in Experimental Watershed Hydrology, Current Perspectives in Contaminant Hydrology and Water Resources Sustainability, Dr. Paul Bradley (Ed.), ISBN: 978-953-51-1046-0, InTech, DOI: 10.5772/55087.

CHAPTER 12

Jian Liu, Dan Liu, and Kai Song, "Evaluation of the Influence Caused by Tunnel Construction on Groundwater Environment: A Case Study of Tongluoshan Tunnel, China," Advances in Materials Science and Engineering, Article ID 149265, in press

INDEX